The Meaning of Proofs

The Meaning of Proofs

Mathematics as Storytelling

Gabriele Lolli

translated by Bonnie McClellan-Broussard
foreword by Matilde Marcolli

The MIT Press
Cambridge, Massachusetts
London, England

The translation of this work has been funded by SEPS Segretariato Europeo per le Pubblicazioni Scientifiche

Via Val d'Aposa 7, 40123 Bologna, Italy

seps@seps.it—www.seps.it

The MIT Press would like to thank the anonymous peer reviewers who provided comments on drafts of this book. The generous work of academic experts is essential for establishing the authority and quality of our publications. We acknowledge with gratitude the contributions of these otherwise uncredited readers.

This book was set in Times New Roman by Westchester Publishing Services. Printed and bound in the United States of America.

Library of Congress Cataloging-in-Publication Data is available.

ISBN: 978-0-262-54426-9

10   9   8   7   6   5   4   3   2   1

Voglio trovare un senso a questa storia . . .
I want to find a meaning in this story . . .
—Vasco Rossi, Italian rock musician

Contents

Foreword

The reader is warned: this is not a typical math popularization book. It is not a collection of useless anecdotes about the lives of mathematicians, peddling dubious genius mythologies. It is also not a magic show of tricks and games, trying to portray mathematics as a form of entertainment for children and young adults.

It is a book about the structure of mathematical writing, and mathematical proofs, but it is also not a book about how to write proofs, of the kind typically designed to expose undergraduate students to rigorous mathematical thinking.

It is a book about *mathematical proofs as a form of narrative*, a philosophical and poetical reflection about why we write mathematics in the narrative form that we currently use.

The narrative structure of mathematics is analyzed at different levels, from the large scale of the "grand narratives" that influence the development of mathematical research over the course of a significant amount of time, usually through the contributions of many authors and many papers, down to the smaller scale of how individual mathematical statements and proofs are formulated and developed within the narrative of individual papers.

Examples of the first kind include Klein's Erlangen program (1872): a programmatic manifesto advocating the study of mathematical objects through their symmetries and transformations. Over time, this point of view became extremely influential in both modern mathematics and theoretical physics, from Noether's theorem on symmetries and conservation laws to category theory and its many applications. Not all of these programmatic grand narratives are successful, as the author shows by discussing Hilbert's proposal for the formalization of set theory, which was shown to be nonrealizable by Gödel's theorem. He also considers contemporary examples such as the Langlands program.

At the level of the structure of individual proofs, one of the nicest parts of the book, in my opinion, is the discussion of how certain structures of mathematical proof are modeled on specific forms of composition in ancient Greek poetry.

This is a book addressed primarily to an intellectually curious public of nonmathematicians, though professional mathematicians will find it interesting too.

The author often relies on literary examples to convey the ideas, rather than on more explicitly mathematical ones. This use of literature as a proxy for mathematics helps make the point that, as creative activities, mathematics and literary composition are ultimately not so different.

The use of metaphors and images in mathematical reasoning and research is also discussed at some length: this is a very important topic because mathematical research makes crucial use of metaphors, and this is something very few people outside of mathematics are aware of. In the author's words, "To do mathematics means to move along different levels with different languages, to use fiction and computation, to be poets and factory workers." This is surely one of the best descriptions of a mathematician's activity that I have come across so far.

The use of "narrative tools" such as visual diagrams or mechanical devices in the context of mathematical proofs is also discussed. This is not just relevant in the historical terms of how the use of such tools was viewed in antiquity, but also in the modern setting with major ongoing research on automated proof verification.

Other topics covered in the book include the evolution of mathematical concepts over time, with historical examples like infinitesimals and infinities, the history of the use of symbols and various forms of logic, and philosophical reflections on the notion of mathematical entities. Lolli's own mathematical research is in the field of logic, and this naturally tends to color the choice of topics in the book, but overall the view of mathematics presented to his readers strikes a good balance between historical and modern, and it manages to convey a sense of the very broad span of different areas of research within the field.

I personally would like to see more books on mathematics for the general public that have this kind of cultural span. I also think that a deeper understanding of the narrative forms of mathematics will be of help to both mathematicians and a broader audience interested in mathematics as a part of our culture.

As the one-sentence back cover blurb in the original Italian edition of this book states: "Poetry says more with fewer words. So does mathematics."

Matilde Marcolli

Preface

In this book we'll be talking a lot about mathematics and above all about proofs. The purpose of the ideas presented here is to demonstrate and convince the reader that the natural way to conceive a proof is to construct it like a narrative, and the natural way to understand it is to listen to it like a narrative. This book was published as part of a series entitled Narrating Mathematics. Narrating mathematics doesn't mean talking about the context or the personalities central to creating and teaching mathematics or what they're doing. Mathematicians "don't know the history of yesterday" (like the partisans' children in the 1958 song "Oltre il ponte," by the Italian author Italo Calvino), not even the biographies of those whose theorems they cite. "Narrating" doesn't even mean summarizing what is already established, beyond debate and inanimate; perhaps that could be described as dissemination. Narrating means talking about the true content of mathematics, explaining it. When standing in front of a painting, how many times have we felt the need for an expert to help us understand why what we're looking at is a work of art? And, after the explanation, we see it from a new perspective. When you ask mathematicians what they do, they don't want to (and can't) answer by listing the results they've found, which wouldn't be understood anyway. However, they would like to explain the meaning of what they do (see appendix, 1). The difficulty lies in the fact that the "meaning" is not easily defined; when you grasp it, you recognize it: as Augustine says of time in his *Confessions, Si nemo ex me querat, scio; si quaerenti explicare velim, nescio,* "If no one asks me, I know; if I want to explain it to someone who asks me, I don't know" (11.14–17). Nevertheless, the meaning of events lies in the stories told about them. We can indeed suspect

> that the true power of historical events lies in their telling. It's only due to their transformation into language that they continue to occur. And, once recorded (even if only as a bit of gossip) they become oddly timeless, inhabiting that cataloged present that passes for time in a library. . . . It is a suggestion, I think, of Schopenhauer . . . that what we remember of our own past depends very largely on what of it we've put our tongue to telling and retelling. It's our words, roughly, we remember; oblivion claims the rest—forgetfulness. Historians make more history than the men they write about, and because we render our experience in universals, experience becomes repetitious (for if events do not repeat, accounts do), and time doubles back in confusion like a hound which has lost the scent. (Gass 1969)

Without a story, events are inert materials, as literary critics who discuss creative imagination point out. In and of themselves, the languages of physics and chemistry don't give us adequate tools to talk about the most interesting aspects of social relationships, the "I," the consciousness, and the content of information (we'll discuss this in section 2.3).

Mathematics, imprisoned within its symbols and images, also says nothing—either to those who invent it or to those who listen to it—if its meaning is not narrated in a story. Mathematicians have no difficulty in answering questions like "What is X?" or "How do you find X?" where X is a concept or the solution to a problem; they answer with the definition, the solution, or the procedure for finding X. If someone asks "What does X mean?" or "What does finding X mean?" mathematicians respond as any human responds when they have to explain the meaning of something: the minute they open their mouths they begin framing a narrative (Mazur 2012, 183).

Rather than a rational animal, the human being is essentially a narrating animal, we express ourselves by telling stories. In the prehistoric world, how did *Homo sapiens* prevail over the other human species that inhabited the various regions of the globe, living together with other similar species of primates? The other human species included, as far as we know, *Homo neanderthalensis* in Europe and western Asia, *Homo erectus* in eastern Asia, *Homo denisova* in Siberia (discovered in 2010), *Homo erectus soloensis* in Java, *Homo floresiensis* (a type of dwarf) on the island of Flores, and in Africa *Homo rudolfensis*, *Homo ergaster*, and *Homo sapiens* (and more have been discovered since the first edition of this book was published). All with brains about the same size as our own, but the Neanderthals, for example, were more robust than the *sapiens*. In the first contact between Neanderthals and *sapiens*, when the latter left East Africa about 100,000 years ago, the former were initially able to drive them back. Between about 70,000 and 30,000 years ago, all these other species disappeared and only *Homo sapiens* remained. Two theories about this decisive event have been proposed and are still being discussed. The theory of interbreeding hypothesizes that species crossbred and merged. The argument against this is that when different species interbreed, the resulting hybrids are usually sterile. The replacement theory, on the other hand, simply notes that the other species were eradicated according to a well-known model from the historical period that followed. In fact, recent DNA studies have shown that there was some interbreeding; some fertile coupling in Europe and the Middle East resulted in between 1 and 4% of our DNA being of Neanderthal origin, while about 6% of the DNA of Aboriginal Australians and Melanesians is of Denisovan origin. However, this doesn't mean that the species merged, only that the differences between them were not enough to entirely prevent fertile interbreeding but sufficient to make such contact rare.

So, how did *Homo sapiens* eliminate the other species? Historian Yuval N. Harari (1976–) speculates that a random mutation changed the language abilities in the *sapiens* brain, causing a cognitive revolution. Language serves to convey information, but what is relevant concerns human beings themselves. They are social animals. For them, cooperation is essential for survival and reproduction. More important than knowing where the lions and bison are is knowing who is honest and who untrustworthy, who are allies, and who is sleeping with whom. Language evolved as a way of chatting, exchanging rumors and gossip (Devlin [2000] also reached this conclusion via a different route). Packs, clans, and groups of cooperative animals based on direct physical familiarity stay in small clusters. Groups of more than one hundred individuals are almost unknown. Now *Homo sapiens* has exceeded this threshold: "The secret was probably the appearance of fiction. . . . Any large-scale human cooperation . . . is rooted in common myths that exist only in peoples' collective imagination. . . . Yet none of these things [religious beliefs, nations, laws] exists outside the stories that people invent and tell one another" (Harari 2014, 32). "The truly unique feature of our language is . . . the ability to transmit information about things that do not exist at all," legends, myths, religions, institutions: the "ability to speak about fictions" (ibid., 29).

Bruce Springsteen (1949–) said stories save lives. In a 2017 conversation with Tom Hanks, as part of New York's Tribeca Film Festival, he remarked that stories were powerful enough to save lives, adding that writers told stories to save themselves.

As we will see, for our ancient ancestors, stories explained life and its conditions, the fate that dominated them. Couldn't mathematicians tell stories to give meaning to their work? If they try to explain the meaning, inevitably a narrative—which combines intentions, goals, projects, desires, knowledge, actions, conventions, and interpretations—begins. These are their driving passions.

Mathematics, produced by a narrator, uses the varied expressive forms inherited from cultural tradition; as varied as those of literature that range from novels to short stories to poetry. The founding mathematical programs of the past—Klein's, or Hilbert's—or of the present—Langlands's—recall the founding myths (chapter 1) and guide research. Every single proof is the story of an adventure, a journey into an unknown land to open a new, connected route (section 2.2). Once the road has been opened, you then improve it. The first time perhaps it's long and bumpy, then you discover the shortcuts; or vice versa, the first route is quick but mathematicians prefer those which ascend to the high peaks and offer broad panoramas (section 2.1).

The fairy tale is a particularly relevant model for mathematics (section 2.4). Fairy tales are a spiritual guide for humanity, along with "fantastic" stories and poetry, because they are a repository of invented worlds—worlds parallel to or interwoven with the one in which we're living—that have no fear of incongruencies. These possible worlds reveal the way in which human beings have lived, thought, hoped, understood, or disputed with the world around them.

As every fairy tale offers a narrative structure in which new characters can be inserted into the recurring forms of the genre in original ways (Rapunzel, Snow White, Cinderella . . .), so each new abstract concept is the protagonist of a different theory supported by the general techniques of mathematical reasoning. When they come up with concepts that are absurdly contrary to common sense—they could be points at infinity, or actual infinities (section 3.3)—mathematicians have no hesitation in cooling metaphysical enthusiasms by minimizing or demystifying: then they call them "fictitious," or fictions.

A common critical opinion is that "the narrative structures" within which fairy tales take place "exist on their own as geometric figures or Platonic ideas or abstract archetypes" (Calvino 1995, 1694); not only that, but the narrative structures preceded and perhaps prefigured the tales. We too should mature seamlessly from the age of fairy tales to the age of scientific knowledge. If you study the evolution of Western civilization, you recognize that there is more than just an analogy between literature and mathematics; there is a direct influence. From cosmological myths to Homeric epic, lyric poetry, Greek tragedy, rhetoric, and history, the Greeks refined and perfected language and reasoning to the point of codifying logic (chapter 6). This path can be traced straight to Euclid's proofs, where the first logical rules can be seen at work. Their roots in poetry and rhetoric are both evident and documentable (chapter 7).

References to mathematical concepts and symbols that, although available to the readers of this series, might not be familiar to everyone, are gathered in the appendix, more or less in the order in which they appear in the text.

1

Great Stories

Ognuno è un cantastoria.
Everyone is a storyteller.
—Mario Castellacci, Italian lyricist

Physics lends itself more readily to considering the relationship between theories and stories. Basically, cosmological myths, early natural philosophy, and different versions of pre-Socratic metaphysics are all precursors or original forms of physics theories. Just think of Anaximander of Miletus who, around 610 BCE, proposed that the origin of all things was infinity (ἄπειρον), which separated pairs of opposites through rotary movement. These theories naturally culminated in the atomism of Lucretius (94–55 BCE), and they are all good stories. They are the stories which taught the lesson modern science has learned and pursued: the real world is the fantastical one, unrecognizable to the senses.

1.1 Hamlet's mill

The origins arise in the distant past. In the late Neolithic, in the fifth millennium, between Chaldea, Egypt, and India there developed

the colossal features of an archaic astronomy, which fixed the course of the planets, which gave its name to the constellations of the zodiac,

which created the astronomical universe—and with it the cosmos—
which we find already at hand when writing first develops, around
4000 BCE. . . . The proportions of Mesopotamia's ziggurats [inspi-
ration for the story of the tower of Babel] bear witness to their
knowledge of how to calculate astral time, as does the arrangement
of megaliths at Stonehenge.

Five times in the course of eight years it happens that Venus (the
morning star) rises just before sunrise (a solemn moment in many
civilizations). Now, those five points (where Venus rises) marked on
the constellations' arc, and joined according to the order of their
succession, reveal that they form a perfect pentagram (i.e., a five-
pointed star). This seems like a gift from the gods to humankind, a
way of revealing themselves. Thus the Pythagoreans said: Aphro-
dite has revealed herself in the sign of Five. . . . But what intensity
of attention and memory it must have taken over the course of eight
years to fix those five blinking glimmers of the planet in the five positions. (Calvino 1995, 2085–2086, review of de Santillana 1985)

These meticulous and precise calculations have given rise to stories throughout the inhabited world. Back then, astronomy and
stories were the same thing. The protagonists were gods, planets.

There is an autochthonous American folk tale, which is still told to
children: Rip Van Winkle, who went to chop wood in the forest and
fell asleep. And here he dreamed of being on the deck of the ship
of Hendrick Hudson, the great ancestor navigator, and of seeing
him and his companions playing at ninepins with cannon balls. He
watched and watched, and then he woke up, and went back to the vil-
lage where he found that nobody recognized him because 300 years
had passed. It is a fairy tale that seems to have been born among
New York's first settlers, but when the ethnologist discovers that
there are versions in every culture that go back thousands of years,
and become clearer the further back you go, one has to conclude that
this is a great astronomical myth: the gods who launch their thunder-
ing spheres on the deck of the sky. And, to the contemplator absorbed
in calculations, a thousand years become like a minute. This was a
way of overcoming Fate; passing from time to eternity, making of
Necessity a free creation. It is the method of the great mathematicians
of all time. (de Santillana 1985, 327–328)

He continues:

> This is linked to a Mayan custom in the Yucatán, who had in each
> of their cities a large ball court, called the "Star Ball Court." This is
> where, on certain solemn occasions, actors depicting the gods played
> according to strict rules, passing the ball through rings arranged in
> the middle of the field: and their names indicated only certain dates
> on the astronomical calendar.

Astronomy is the primary source of myth, but mathematics comes before the myth and number is its interpretive code. As Calvino points out in his essay on the work of science historian Giorgio de Santillana (1901–1974) and ethnologist Herta von Dechend (1915–2001), the keys to all myths

> are the regularities of zodiacal time and its irreversible changes over a
> very long period (precession of the equinoxes) due to the inclination
> of the ecliptic relative to the equator. Humanity carries with it a distant
> memory of celestial movements, so much so that all mythologies retain
> the trace of events that occur every 2,400 years, such as the change
> in the zodiac sign in which the sun is located at the equinox; not only
> that, but almost equally ancient is the forecast that the continuous,
> ponderously slow movement of the firmament is brought together in
> an immense cycle or Great Year (26,900 of our years). The twilight of
> the gods recorded or foreseen in various mythologies is connected to
> these astronomical recurrences; sagas and poems celebrate the end
> of time and the beginning of new eras (Calvino 1995, 2087).

The myth par excellence is that of Hamlet's mill, reconstructed by de Santillana and von Dechend, starting from Shakespeare's *Hamlet*, and going back to the sources of the legend of Hamlet in Danish chronicles and Nordic mythologies (Amlethus son of Orvendel, Amleth, Amlóði). Then, involving the Dogon from Africa, Hinduism, the Aztecs, along with Greek and Latin authors, de Santillana and von Dechend trace the emergence of one of the first philosophical problems: the idea of an ordered cosmos whose norms are devastated by a physical and moral catastrophe in response to which arises the aspiration of finding harmony.

Amlodhi was identified, in the crude and vivid imagery of the Norse, by the ownership of a fabled mill which, in his own time, ground out peace and plenty. Later, in decaying times, it ground out salt; and now finally, having landed at the bottom of the sea, it is grinding rock and sand [in some southern Italian dialects—where we still find remnants of languages from Greek and early native civilizations, before the Roman conquest, e.g., in Calabria—older people still use phrases like *stannu carendu i mulinu* and *jetta petri i mulinu* "it's raining millstones," or better yet "it's throwing millstones" to describe a raging tempest], creating a vast whirlpool, the Maelstrom ["the grinding stream" from the Dutch verb *malen*, "to grind"], which is supposed to be a way to the land of the dead. This imagery stands . . . for an astronomical process, the secular shifting of the sun through the signs of the zodiac which determines world-ages, each numbering thousands of years. Each age brings a World Era, a Twilight of the Gods. Great structures collapse; pillars topple which supported the great fabric; floods and cataclysms herald the shaping of a new world . . . the measures of a new world had to be procured from the depths of the celestial ocean and tuned with the measures from above, dictated by the "Seven Sages," as they are often cryptically mentioned in India and elsewhere. They turn out to be the Seven Stars of Ursa, which are normative in all cosmological alignments on the starry sphere. (de Santillana and von Dechend 1977, 2–3)

In these stories Hamlet is always presented as a hesitant simpleton, yet behind his apparent inability falls the shadow of Fate and the distant traces of a necessity, to which not even the gods—with their power to dispense good and evil—are immune. Hamlet should not be seen as an antihero, but rather as a dispenser of justice.

With the advent of writing, the identification of the human mind with the movements of the heavens is lost. Plato (ca. 428–ca. 348 BCE) is the last of the archaic and the first of the modern philosophers. He accepted myth as thought in the form of a story, both as a way to showcase his own doctrines as well as to overcome the limits of rational investigation: the myths of the Demiurge, the cave, the cosmic cycles, the golden age, the immortality of the soul, and others. With Aristotle (384–322) cosmic wisdom fades away. In the previously cited essay on de Santillana and von Dechend, Calvino expresses the opinion that the archaic philosophers were the true representatives of a scientific spirit, as opposed to contemporary humans, who believe they can use natural forces as they please and therefore participate in something like magical thinking (Calvino 1995, 2089).

If myths are stories intertwined with early astronomical science, conversely theories in physics, especially contemporary ones, sometimes make us suspect that they are stories. It's no surprise to hear that contemporary physicists, who know how to read and write, turn to literature to fill gaps and shortcomings, or simply out of dissatisfaction with scientific explanations: science (quantum gravity) shows how time disappears from the fundamental structure of the world; "loop quantum gravity shows that it is possible to write a coherent theory without fundamental space and time—and that it can be used to make qualitative predictions" (Rovelli 2018, 79).

Consequently, "the human condition of being in time is not analyzed solely by way of scientific laws, but it is explored by philosophical concepts and above all through narrative categories and poetry that distill and coagulate, transfuse and combine understanding and intuition, knowledge and feeling" (Rovelli, quoted in Bricco 2017).

Theoretical mathematics doesn't seem like a system of stories, although its growth has been marked by human time. However, the obvious paradox is that it is commonly considered eternal and timeless, a view justified by the way mathematicians talk about their discipline.

1.2 The universe of sets

In mathematics, the only story comparable to the ambitious "theories of everything" found in physics is set theory. Set theory recounts how the entire universe of mathematics was born. The grand story of set theory is similar to a cosmogony. There is something miraculous about the way the universe is formed because it is the structuring of nothing. We begin from nothing, the empty set $\emptyset$, which is to say a set that contains no elements. By repeatedly performing two "set-forming" operations, the pair $\{x, y\}$ and the union $\cup x$, we obtain all of the hereditarily finite sets. Among these, there are groups that can be ordered in a sort of spinal column and to which the properties of natural numbers are attributed: $\emptyset$, $\{\emptyset\}$, $\{\emptyset, \{\emptyset\}\}$, $\{\emptyset, \{\emptyset\}, \{\emptyset, \{\emptyset\}\}\}$, . . . (see appendix, 5).

However, without infinite sets, the universe would be too small. There would be no big bang that produced it all, just the fizzle of the first fuse. It's necessary to imagine a series of big bangs close together, as some physicists have done: the first makes an infinite set appear, $\mathbb{N}$, or ω, or $\aleph_0$ (every discipline has different names, depending on the function they perform—natural numbers, finite ordinals, the cardinality of $\mathbb{N}$—but they are all views of the same thing), and the first is followed by other larger ones, $\aleph_n$. Two actions happen to create the endless iteration of the big bang. On the one hand, since it is not possible to construct all of the subsets of an infinite set in real time, explicitly indicating the elements, another phenomenon is necessary, understandably called "power": the power of x, $\mathcal{P}(x)$, is the set of all of the subsets of x, and it is an infinite set that is yet larger than x, even if x is infinite (Cantor's theorem; see section 2.1).

The significance of $\mathcal{P}(x)$ is clear, also through analogy with the finite case: if x is finite and has n elements, its subsets are countable, and they are 2^n; but the meaning of the power is not evident, because the concept is ambiguous for an infinite x. Talking about *all* is a shortcut, it is an attempt to say the totality, or rather perhaps the aspiration to say the totality; but the majority of subsets are like the dark matter of physics, much more extensive than visible matter. The totality cannot be unanimously fixed. $\mathcal{P}(x)$ has to be approximated, and its meaning is only produced by a plurality of partial results applicable to it. Which part of $\mathcal{P}(x)$ we're able to control is not a given: the finite subsets of x, x itself, x minus finite subsets, of course; however, in order to see, for example, an infinite subset whose complement in x is also infinite, it needs to be defined. That's the case with the subset of the even numbers in $\mathbb{N}$, or that of Fibonacci numbers. Those that can be defined are the only ones we're sure everyone recognizes. And yet, because of Cantor's theorem, we know that all of the subsets are not exhausted, because there are as many definable subsets as there are language formulas: a countable infinity.

We see definable sets indirectly through their definitions, with linguistic tools, and this is the second action. The universe must respect our way of dealing with visible matter. For example, if the elements of an infinite set—elements scattered throughout the universe and even vastly distant from one another—are associated with a functional formula, then there is a set that brings them all together in a single galaxy (axiom of replacement): if, let's say, every $n \in \omega (= \mathbb{N})$ is associated with $\aleph_n$, there is an $\aleph_\omega$ which contains all of the $\aleph_n$. This is the axiom which, together with the axioms of infinity, of the power set, of pairing, and of union, substantially constitutes the ZF theory of Ernst Zermelo (1871–1953) and Adolf A. Fraenkel (1891–1965), with some cosmetic tweaks to make it beautiful, elegant, and minimal. The axiom of replacement is a schema of axioms, that is, one for each functional formula; with this axiom it is possible to prove the existence of each definable subset of a set (which in Zermelo's original theory was an axiom). Finally we add the axiom of restriction (foundation axiom), which says the descents from a set to its elements and the elements of the elements, etc., are always finite.

The universe V has taken shape, that of a cone with the vertex in $\emptyset$, and expanding in growing circles (like Dante's *gironi*): the natural numbers form a set ω and have an extension with $\omega \cup \{\omega\} = \omega + 1, \ldots$ in the infinite ordinals, which are always the backbone of the universe; ordinals measure the stages of generation.

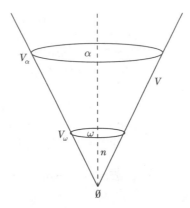

The initial segment of the universe V formed by hereditarily finite sets is a small set which is indicated by V_ω. The symbol V (as if prescient of what it must denote) derives from the reversal of the Λ, used by Giuseppe Peano (1858–1932) to represent the empty set.

Numerical systems come to light through the combined action of power and replacement, not only $\mathbb{N}$ but also the rational, real, and complex numbers, even those which have an infinite power greater than $\mathbb{N}$, greater than the countable. By definition relations are sets of ordered pairs, and it is the same for functions, or operations. And, with relations and functions, structures of any type can be materialized. Mathematical entities have all been created.

This account of the formation of V has two flaws; on the one hand, the story doesn't correspond to the chronology, which we are close enough to be able to document. When mathematicians decided to study infinite sets (thus challenging the excommunication of the actual infinite dating back to the Greeks), they obviously considered them as given, in particular what was before their eyes in the numerical systems they already knew, such as $\mathbb{N}$ or $\mathbb{R}$; the empty set was the last thing they were worried about. The succession of big bangs is a reconstruction after the fact to give an order to and a justification for the axioms, which were chosen pragmatically and with care from the properties most used in proofs for propositions involving sets. The concern was to deal with how we reason, not what exists; which is to say the discourse determines what exists.

On the other hand, although mathematicians are willing to accept that a mathematical structure—let's say a group—is a set with an operation that satisfies certain properties, indeed the generality of the description is an advantage, they cannot imagine why they should conceive the number 3 as {∅, {∅}, {∅, {∅}}} (the reason for this definition is that the order relation between natural numbers coincides with the membership relation ∈, with significant simplifications—in set theory, certainly not in arithmetic). Mathematicians are less likely than physicists to accept that the real world, according to what science teaches, is different from the perceived one, because for them there is only one, the one they've imagined.

Nevertheless, the story isn't realistic as a temporal story or even as a mathematical story; it is an origin myth that is useful as a methodology, to justify the reducibility of all mathematical concepts to sets on the one hand and on the other the continued enrichment of the theory. In fact specialists have begun to get impatient, as they have very little to do, because it seems as if everything relative to the foundation of mathematics has already been done. Naturally, that's aside from the inevitable incompleteness pointed out by logicians and therefore the existence of undecidable propositions, which entails the existence of different models of the universe. In truth, there is no need to appeal to the general theorem of Kurt Gödel (1906–1978) for the incompleteness of set theory, which results from the fact that the power operation is not well defined. In fact, it is not possible to determine the cardinality of $\mathcal{P}(x)$, even if it is indicated by an operation similar to the finite exponential, $2^{card(x)}$; its values are different in different models.

Therefore, far from there being just one universe V, and ZF theory of V, the models of the theory, the universes, are myriad. A theory of the multiverse is currently being developed, not unlike what is happening in physics. Analogies with cosmology suggest that mental restrictions exist in shaping very general theories. On the other hand, the current fashion of the plurality of universes, physical and mathematical, seems to be in tune with the intense passion for digital video games and virtual reality. But why should the universe end like this? After the first big bang, which caused a set to appear containing all the hereditarily finite sets, and after the automatic iteration of the big bang that resulted in the higher cardinalities, why couldn't there be another epoch-making big bang that makes V become a set and allows generation to restart, and iterates it, with other sets whose existence is not demonstrable in ZF and in the expanded theories? And we begin again, calling the upper limit "inaccessible," and placing ourselves right beyond the limit, since we cannot reach it. And the leap beyond the sets whose existence is demonstrable in the new theories is repeated. It's called the large cardinals program, initially suggested by Gödel in 1946.

Ultimately, in this history of the formation of the universe of sets, it will be understood that philosophical intents and objectives prevail, as the myths did in other areas. In fact, in this case we're not talking about real people—despite the fact that theories are formed from their ideas and decisions. The acronym ZF is two initials linked to two names, but Thoralf Skolem (1887–1963) has been left out. Few know the fascinating page of Zermelo in which he explains how he found the axioms while studying the proofs that were done before the axioms were thought of. Heated and violent disputes have accompanied the acceptance, or refusal, of ways of reasoning about the infinite, for example using the axiom of choice. The disputes were not sterile ends in themselves, but produced alternative developments, although not all with the same following and weight. Physics speaks of matter moved by laws to which consciousness is foreign; its discriminating criterion is that of observations of the matter, decisive with respect to any conflicting opinions between physicists. This is not the case in mathematics; as in the old adage "it takes money to make money," those with the most richly developed and well-illustrated construction come out on top. To allow time to enter into the equation, Rovelli suggests that we turn to literature, but time is already present in mathematics, and it always has been—with the exception of the Babylonians and the Egyptians, etc., back when mathematics was only a recipe book of problems and solutions. Proofs introduce motion (section 3.2).

1.3 Groups and geometries

There are other beautiful stories in contemporary mathematics known as programs. Perhaps the first to be called such was the Erlangen program (1872) of Felix Klein (1849–1925). This program unified the study of new geometries based on transformations that preserve the essential properties of the characteristic figures of each geometry. Klein suggested the group concept for classifying geometries.

A group is one of the simplest structures; it has only one associative binary operation, with a neutral element and such that each element has an inverse.

In Klein's time, the most interesting geometries were projective, affine, Euclidean, and non-Euclidean; but there were few known groups, basically the group of permutations of n letters, the group of regular polygon symmetries, and a few others; the terminology was not established, much less the theory. The definition of abstract groups was made by Richard Dedekind (1831–1916) in his little-known lectures of 1856–1858 on the work of Évariste Galois (1811–1832) regarding the solutions of algebraic equations (we'll be discussing Galois groups a bit later). In 1854, Arthur Cayley (1821–1895) proposed a system of axioms for finite groups.

The known groups were mainly transformation groups and symmetry groups. Klein had the idea that each geometry was characterized by a group of transformations and that the object of each geometry was the study of the invariant properties with respect to this group. He even thought of defining a geometry as the study of the invariant properties of figures (that is, those properties that do not change) when subjected to a "group of transformations."

The transformations that characterize Euclidean geometry are rigid movements, isometries. Affine geometry is the study of properties and relations that are invariant with respect to the affine group (the group of transformations which are the composition of a linear transformation and a translation); they send straight lines to straight lines and parallel lines to parallel lines, while they do not necessarily maintain the distances and angles. Projective transformations are those that transform aligned points to aligned points; only the properties of incidence and the cross-ratio are preserved in projective transformations (see appendix, 6). The expanding hierarchy of inclusion is therefore: projective, affine, Euclidean.

Klein showed that, starting from the group of projective transformations which preserve a particular conic curve, one reaches hyperbolic or Lobachevskian geometry, one of the non-Euclidean geometries. Riemannian geometry is not covered by the Erlangen program.

Group theory thus came to have a geometric meaning, increasing its presence and prestige. Later it also found a place in physics in the program initiated by Emmy Noether (1882–1935) to characterize the laws of conservation in physics based on invariance relative to the symmetries of a physical system. Momentum is conserved if the system has a symmetry for translations, angular momentum for systems with radial symmetry, invariants for rotations, energy for temporal symmetries, etc. Klein's program was exhausted in its formulation, because the concepts to be combined and unified were given. Then, of course, it rippled through the conceptual organization of mathematics and didactics like a shock wave. The realization and awareness of groups of transformation in physics was progressive, starting with Noether's 1918 theorem which states that every Lagrangian symmetry of a physical system has a corresponding conserved quantity. As proof that mathematicians don't know the history of yesterday, in an article for the *New York Times* published on March 26, 2012, Natalie Angier wrote about a survey done some time ago by Dave Goldberg which aimed to evaluate the popularity of Emmy Noether among dozens of his colleagues and physics students. The investigation revealed how few people knew about her and her work.

1.4   Hilbert and infinity

The program of David Hilbert (1862–1943), developed in the 1920s, offers a compelling story about the infinite, disappointing or even tragic in its conclusions. Unlike Paul Klee's *Angelus Novus*, it is not advancing with its face turned toward the past, which could be identified with the universe of sets; nor is it exhausted in the present, as one might say of the Erlangen program. When conceived, Hilbert's program was a real project, aimed at the future, in harmony with the word "program."

The widespread use of the word "program" lately seems to be due to the influence of post-neopositivist philosophy of science; Hilbert called what he was doing "metamathematics," and conceived it as the construction of a new discipline.

Hilbert said that he intended to restore certainty to a mathematics plagued by arguments about infinity and the threat of antinomies, as well as to defend it from the attacks of the intuitionists, Luitzen Egbertus Jan Brouwer (1881–1966) and Hermann Weyl (1868–1927), who wanted to reformulate mathematics, mutilating it, using a logic weaker than classical logic which was, in their opinion, more suitable for a finite creative subject (see appendix, 10).

Hilbert proposed: (a) to formalize the theories of infinity, analysis, or set theory; (b) to demonstrate their noncontradiction in a progressive way, from arithmetic without quantifiers to Peano's arithmetic, to real number theory, to the whole of set theory; (c) to demonstrate it using absolutely reliable elementary methods. The proof of noncontradiction with combinatorial methods, finitary methods as they are called, would have permitted an unexpected conclusion: the conservativeness of the theories of infinity on elementary mathematics, that is, the fact that every numerical relationship demonstrated with the help of infinity was true and already provable with elementary methods. At the end of the story, the danger of infinity would no longer have been a nightmare and the concept would have been downgraded to a mere tool (see Lolli 2016a; we'll discuss this further in section 3.3).

More generally, metamathematics had to confront and resolve the epistemological questions emerging from the study of proofs in a correct and secure way—"the problem of the *solvability in principle of every mathematical question*, the problem of the subsequent *checkability* of the results of a mathematical investigation, the question of a *criterion of simplicity* for mathematical proofs, the question of the relationship between *content and formalism* in mathematics and logic, and finally the problem of the *decidability* of a mathematical question in a finite number of operations" (Hilbert 1918; English translation in Ewald 1996, 2:1113). The discipline was constructed, with useful repercussions; however, the main goal proved impossible. In reality the script was never written; several indexes appeared, gradually more detailed, and some test pieces. The story's ending clashed with another, more convincing story: incompleteness. The incompleteness promulgated by Gödel was more convincing, first and foremost because it was a statement of the facts rather than a program. Few studied the proof immediately, but many were convinced by Gödel's brief explanation at Königsberg in 1930 (von Neumann before anyone else) that noncontradiction implied the deductive incompleteness of arithmetic, and therefore it was not demonstrable because, contextually, incompleteness was one of the undecidable statements. In addition, Gödel presented his result (especially to Zermelo who did not accept it) with modesty, almost minimizing it. He emphasized that it depended on having considered restricted proof methods (first-order logic); it then seemed plausible, considering weak premises and a strong conclusion (completeness), that it was impossible. On the contrary, Hilbert, using even more restricted methods, set his sights on opening the doors of what he called Cantor's Paradise.

What Hilbert had truly feared, and the reason he tried to stymie Brouwer, was the mutilation of mathematics. In the end, he needn't have worried; mathematicians loved their creation too much to worry about the warnings and prohibitions Brouwer wanted to impose. They would even have had to learn another type of logic for infinite constructions. But Hilbert, who was broadly liberal when it came to the axiomatic method, was a true absolutist with respect to classical logic. Intuitionism has not particularly benefited from the legitimacy indirectly conferred on it by the failure of Hilbert's program; but a different, interesting, constructive (albeit niche) mathematics and a new logic have certainly developed. Had the flawed story come to an end, not only would we have been deprived of this new mathematics, we also would have lost important contributions to logic, for example those that came from the reinterpretation of intuitionistic logic in terms of the logic of problems, by Andrzej Grzegorczyk (1922–2014), or the semantics of possible worlds for modal and temporal logic in the work of Saul Kripke (1940–) (Grzegorczyk 1964; Kripke 1963).

1.5   The Langlands program

Another program currently under way, which we will briefly discuss, is the Langlands program. Robert Langlands (1936–) is a Canadian-born mathematician who taught at Yale and was a professor at Princeton's Institute for Advanced Study. His program gathers together important stories, true stories that are shaping future mathematics. It also has foundational ambitions, different from those of Hilbert's era, yet not dissimilar from the Erlangen program. The latter linked different areas such as group theory and geometry. The Langlands program aims to unify the divergent dispersion caused by the proliferation of specializations that, due to the internal drive toward growth, seem no longer related to each other. However, the objective is achieved through theorems that deal with the relationships between the most advanced and specialized concepts of the disciplines considered. This combination is not only surprising but also a source of unexpected discoveries. We're on the cutting edge of research.

Fortunately one of the most respected and creative mathematicians working on the program, Edward Frenkel (1968–), had the courage to try to explain it, so that one can utilize his competence as an insider. We refer to Frenkel's exposition (2013) for further explanations. Here we briefly touch on a few points for those readers who are not afraid of a deluge of mathematical concepts they've never heard of (see appendix, 7, 8, 9).

Although the subtitle of Frenkel's book is "The Heart of Hidden Reality," his approach is compatible with ours, given that he states that the idea behind the Langlands program is "bringing together the theories coming from these diverse fields and realizing that they *are all part of a single narrative*" (Frenkel 2013, 71, emphasis mine).

Readers will have no difficulty in accepting that numbers and forms are presented as different things, and they are, in the sense that they require different skills from the mathematicians who study them—who can generally be classified as analytical or geometric. Klein himself introduced the first characterization of types of mathematicians. This was during the years when specialization had begun to make communication and collaboration between mathematicians working in different disciplines difficult. But in reality numbers and figures have been linked since the time of the Greeks, who had a geometric algebra; by this we mean that they knew how to demonstrate arithmetic properties with geometric methods (e.g., the commutativity of multiplication, *Elements* VII.16), geometric versions of what, for us, are algebraic formulas (the square of a binomial, *Elements* II.4), and how to solve equations (the term "geometric algebra" is controversial; it was supported by H. G. Zeuthen (1839–1929) for Euclid and especially for

Apollonius of Perga (262–190 BCE); B. L. van der Waerden (1903–1996) called Apollonius a virtuoso of geometric algebra). Later, the link between numbers and figures became even closer with the oxymoron of analytic geometry. When groups started to appear in the narrative, they were linked to geometry, as we mentioned in our discussion of the Erlangen program. Consequently geometry came to rely on algebra rather than on numbers.

The groups under consideration in the Langlands program are Galois groups. Initially, these were only a technique for studying the solutions to algebraic equations and the possibility of expressing them through roots (in addition to the four operations). They are symmetry groups of numeric fields obtained by adding the solutions of an equation. From Galois's work we know that an equation admits a formula to express solutions for radicals if and only if the group has a particular property (it is a solvable group; and fifth-degree equations are associated with an unsolvable group).

Langlands's first crazy idea was to link the theory of Galois groups of finite fields modulo p with harmonic analysis. Let us consider equations of two variables and their modulo p solutions, for p prime. The number of solutions is close to p, the difference is a_p. In 1954, Martin Eichler (1912–1992) found that, for a cubic equation, these seemingly random a_p deficits are the coefficients of a generating function (a series) that has a symmetry; it is invariant relative to the symmetries of the complex unitary disc group. This function is called a modular form. The function is like the sine for Fourier analysis, a basic wave, a harmonic; in fact it is studied in harmonic analysis.

The first step of the Langlands program is a generalization of Eichler's result, which Langlands expressed in a letter to André Weil (1906–1998). It has been widely distributed and led to the formulation of the Shimura-Taniyama-Weil conjecture: for every cubic equation, the number of modulo prime solutions are the coefficients of a modular form.

The conjecture, with the leap to all cubic equations, was very courageous. It implies the proof of Fermat's theorem, which Andrew

Wiles (1963–) proved in 1994. A first generalization regards the correspondence between representations of the Galois group and the automorphic functions. Then Weil introduces a third term, giving a geometric version of the program; it is now a question of linking results from three sectors:

number theory	curves on finite fields	Riemann surfaces

Weil describes his work within the program as that of a linguist who has to decipher a trilingual text in the three columns, currently having only a few fragments; years of work have led to some additions to the dictionary. Each individual case requires a specific and demanding demonstration.

With three columns, it is necessary to have the analogues of the others in each one. The fundamental group for the Riemann surfaces was invented by Vladimir Drinfeld (1954–), one of the program's pioneers. For the correspondents of the automorphic functions, a generalization of the sheaf-based function concept was considered. That concept was developed by Alexander Grothendieck (1928–2014). The term *faisceau* or "sheaf" was first used in 1946 by Jean Leray (1908–1998) in his research on algebraic geometry.

In vector spaces it is possible to introduce operations similar to those of addition and multiplication on the natural numbers; vector spaces are not simple objects like numbers, there are functions between one and the other, morphisms; categorical language becomes indispensable. When the stories become too complicated, an appropriate language needs to be found. The traditional concept of function appears archaic.

With things like a sphere or a torus, a sheaf is a way of associating a vector space, rather than a number, with every point, with coherence conditions on the fibers which there is no need to remember and which are well expressed with the commutativity of morphism diagrams.

A further step in the generalization of the program was the insertion of a fourth column with quantum physics. This involves the concepts of electromagnetic duality and mirror symmetry in the geometric program. Without going into details, the program is enhanced with a new column.

number theory	curves on finite fields	Riemann surfaces	quantum physics

The electromagnetic field equations of James Clerk Maxwell (1831–1879) are invariant in a vacuum for the exchange of the two fields, electric and magnetic. The exchange is a symmetry of the equations and is called electromagnetic duality. At small distances and high energies, the behavior of the two fields is described by quantum field theory. The two fields are produced by photons, which interact with other elementary particles. To exhibit a duality in quantum electromagnetism, one would have to exchange electrons and magnets, but magnets have two poles that cannot be separated; in 1931 Paul Dirac (1902–1984) introduced magnetic monopoles as a field singularity. We still don't know if they exist; however, by assuming them, a theory that exhibits an electromagnetic duality can be constructed. It was initially thought that since fermions and bosons are two very different types of particles, an electromagnetic duality should have preserved a distinction between the two particles. However, in the 1970s the idea emerged that a special type of symmetry could exchange them; it was called supersymmetry. Supersymmetry theories have several advantages; quantum electromagnetism is not supersymmetric but has a supersymmetric extension. While we don't know if an electromagnetic duality really exists, it manifests itself in a supersymmetric extension of the theory.

To enhance the picture, we must also consider the nuclear forces: the strong nuclear force (which holds quarks inside the elementary particles) and the weak nuclear force (which is responsible for various transformations). Theories of electromagnetism, for both strong and weak nuclear forces, all fall under the category known as gauge theory. Each is based on a group, called a gauge group.

Electromagnetism is the simplest of gauge theories, and its gauge group is the circular group, that is, the group of circular rotations, or of the circle of radius 1 in the complex field. This is a commutative group. In the 1970s, it was also discovered that there is an analogue of electromagnetic duality between noncommutative gauge theories. If we consider a generalization of electromagnetism in which the circular group is replaced by a noncommutative group, we obtain theories that accurately describe the strong and weak nuclear forces.

If the gauge group of this theory is G, the dual theory will have a dual group which is referred to as ^LG, well known and studied within the Langlands program.

Explaining how to proceed from here is too difficult (see chapters 16 and 17 in Frenkel 2013). Let's just remember that the fourth column was discovered to be linked with the previous ones; for example, not only with the Langlands dual group ^LG but also with the Kac-Moody algebras studied by Frenkel.

It is evident that in this story there are, as we go forward, remarkable new examples of what is called the unreasonable effectiveness of mathematics. It happens quite often that concepts, results, and theories developed by mathematicians with no applicable purpose find an essential place in physics research; the classic example is Apollonius of Perga's conics. In this case we have to recognize that the interaction between mathematicians and physicists is intensive, to the point that the two types of researchers become indistinguishable. One of those who has contributed and is still contributing the most is Edward Witten (1951–), who was the first physicist to receive the Fields Medal in mathematics in 1990.

The reader may suspect that the word "stories" is being used as a metaphor to talk about the programs described, and in a sense that's true. From these grand stories we must descend from the stratosphere to more welcoming climes: simple stories, less drawn-out and with fewer prerequisites, but ones in which the meaning is perhaps more hidden. The programs have a historical and social component that is superficially understandable. In physics it's thought that there is no need to invent meaning; physics is the description of nature, and that is enough to distinguish the true from the false. However, only the experimental consequences of the theories can be verified. It was this that inspired Hilbert when he suggested we would do well to worry only about the truth of numerical equations, and to ignore the meaning of infinity by eliminating the question through a proof of noncontradiction. Instead, the failure of his program pushes us not to put the issue of the infinite aside but to continue to work on it. Physical theories are in themselves pure mathematics, completely involved with infinity, and they are also searching for a meaning. To find it, we must consider the delicate and diffuse network that underlies mathematics, represented by its proofs.

2

The Meaning of Proofs

Your fatal habit of looking at everything from the point of view of a story instead of as a scientific exercise has ruined what might have been an instructive and even classical series of proofs.

—Arthur Conan Doyle

For those readers who aren't familiar with examples of proofs or don't have them at hand, we've included a few below, each quite different from the others. They're challenging but not difficult and, with just a bit of concentration, can be understood. The selection is almost infinite, so it wasn't difficult to choose. Readers who are familiar with mathematics can refer to those they know, and the better they understand them the better they will understand the text. The ability to call to mind a variety of notable examples makes it possible to recognize strategies, moves, and techniques that structure the proofs at work in a concrete way—regardless of languages and styles that have changed over the centuries. They are still changing, as we will see, from the primordial beginnings of a theory to its maturity. Those presented are not paradigmatic; it is precisely their diversity that proves there is no model, no template for proofs. The examples serve in place of a general definition of proofs, which doesn't exist.

2.1   A mixed bag of proofs

Here are the promised proofs.

1. Euclid, of whom we only know that he was active in Alexandria under the reign of Ptolemy I of Egypt (323–283 BCE), and therefore between the time of Plato and that of Archimedes, composed, in addition to other works, the first collection of *Elements* that has been passed down to us (see appendix, 3).

Proposition III.31 of Euclid's *Elements* deals with angles with a vertex on a circle that are subtended by an arc that is either a semicircle, less than a semicircle, or greater than a semicircle. Let's just look at the case of the semicircle, as shown in the following figure, with the vertex at *A*, in a circle with center *E*. As to the other two cases, the reader can easily imagine the conclusion. So we take the liberty to formulate the proposition like this:

Euclid   Proposition III.31: If an angle with a vertex on a circle is subtended by a semicircle, then it is a right angle.

The following figures are taken from Heath (in Euclid 1956, 2:62) which simplifies that of Euclid in the proof of III.31 by eliminating the elements which are useful in demonstrating the other two cases in the proposition. In reality Thomas L. Heath (1861–1940), like Euclid, is looking at a single figure, the last one in our discussion, which we have separated into a series of frames to help the reader see the individual steps of the proof. In addition we've used dashed lines for those elements not given in the hypotheses but added over the course of the proof. Euclid generally used only one figure for the proofs, a figure with all of the elements, which actually appear progressively, already in place. The order in which the figure was constructed from the initial data is meant to be followed by reading the text.

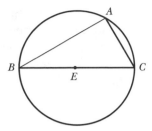

For the proof, a segment *EA* is "produced" which creates a triangle *AEC*, isosceles since it has two equal sides, *AE* and *EC*—both radii of the circle—and another triangle *ABE* also isosceles for the same reason. To "produce" means to construct an element not given in the figure; in particular Euclid's Postulate 2 allows a line to be extended [προσέκβαλλω] (at other times Euclid uses "draw" or other verbs). Postulate 1 allows the drawing of a segment that joins two points.

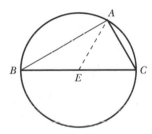

If you imagine all the lines and points of a plane drawn, it becomes uniformly black. The mathematician's art is to "switch them on" one at a time, when they are needed. Producing a line is an act that breaks the functional fixity that tends to show only the elements of the figure with the data of the problem. Creativity, a rare commodity, is manifested in this initiative, since in teaching mathematics we always insist that students must learn to use only what the definitions say, neither more nor less, and rightly so. But in a proposition only the problem is given, not the instruments, which are to be chosen from those indicated in the postulates. In this case producing a segment that makes isosceles triangles appear is an idea that could be a gamble, or an act of desperation. We do it because at least we do know something about the angles of isosceles triangles: the angles at the base are equal (Proposition I.5). Better than just looking at the figure and waiting for inspiration.

In the triangle AEC the sum of the angles $E\hat{A}C$ and $E\hat{C}A$ is the angle $B\hat{E}A$, thanks to the theorem that the sum of the internal angles of a triangle is two right angles (the straight angle in E), as demonstrated in Proposition I.32. Similarly, in the triangle ABE the sum of the angles $A\hat{B}E$ and $B\hat{A}E$ is equal to the angle $A\hat{E}C$.

The sum of the four angles $E\hat{A}C$, $E\hat{C}A$, $A\hat{B}E$, and $B\hat{A}E$ is therefore equal to two right angles, the straight angle of vertex E and sides BE and EC. But the sum of the four addends equals the sum of two times $E\hat{A}C$ and two times $B\hat{A}E$. Therefore, the sum of $E\hat{A}C$ and $B\hat{A}E$ should be equal to a right angle. To use a bastard but perhaps effective notation, one could say that if $2\alpha + 2\beta = 180$ then $\alpha + \beta = 90$.

We say "should" because Euclid proceeds in a different way. The reason is that, among the results already proved, he doesn't yet have a theorem that talks about four angles and their sum (except in the case of quadrilaterals). Euclid's *Elements* is a story that develops with rigid inexorability, as if it were necessary, while it is Euclid's own creation, refined and elegant. Euclid could have proved a theorem that deals with the four angles, useful for concluding the argument geometrically, with his methods. But it wasn't necessary. The reasoning followed is sufficient for the conclusion without involving other properties, at the small cost of producing something new.

Euclid produces an extension BF of BA, making use of Postulate 2.

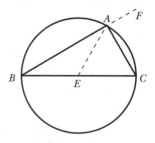

Referring now to the triangle ABC, the angle $C\hat{A}F$, again relying on the same property regarding the sum of the internal angles of a triangle, is equal to the sum of $A\hat{B}C$ and $B\hat{C}A$. But we know from the previous observations that this sum is equal to that of $B\hat{A}E$ and $E\hat{C}A$, that is, to $B\hat{A}C$. Therefore, the two adjacent angles $C\hat{A}F$ and $B\hat{A}C$ are equal to each other. Definition I.10 of a right angle establishes that when a straight line meets another straight line and forms adjacent angles equal to one another, both angles are right angles and the incident line is perpendicular. Therefore $C\hat{A}F$ is a right angle.   □

We will talk more about Euclid's proofs in chapters 4 and 7.

2. Jumping ahead about 2,200 years, we arrive in the era in which Georg Cantor (1845–1918) began discovering a new area of research, which was not yet called set theory. Within mathematical analysis, following the definitions of the real numbers given by various mathematicians in and around 1872, some questions arose about this object, now that it was finally well defined. Richard Dedekind (1831–1916) observed that "the straight line L is infinitely richer in point-individuals than the domain R of rational numbers in number-individuals" (Ewald 1996, 2:770), and Cantor, in a letter of November 1873, says that he believes, in confirmation, that it's not possible to make a correspondence between the natural and real numbers, unlike that which is possible between natural numbers and rational numbers (by "correspondence" Cantor means a one-to-one correspondence; see appendix 4, 5). Dedekind replies that he agrees, but that he has no idea how to prove it. Cantor found the proof in 1873 (he published it in 1874). He could then declare to Dedekind: "I infer that there exist essential differences among the totalities and value-sets that I was until recently unable to fathom" (ibid., 2:846). (The correspondence between Cantor and Dedekind, from which the discussions that accompanied the formation of new ideas and theories can be followed—unfortunately something that we can't do with Euclid—can be found in Cantor and Dedekind 1937; English translation in Ewald 1996, 2:843–878.)

Cantor presented a second proof in 1891. We give them both, to show how the proofs change with the development of appropriate techniques, becoming simpler and more direct, and because the second is much more general, and deals not only with real numbers but with a fundamental concept of abstract set theory, outlined after a decade of intense work. The first, uncertain and arduous proofs can also be illustrated with another instructive example from 1877, that of the equivalence between the square and the side (or between $\mathbb{R}^n$ and $\mathbb{R}$). Cantor's two results from the 1870s are the ones that ignited the fire of infinity theory. In examining the proofs we see that at the beginning the properties necessary to obtain the desired results were taken as a given, when in fact later they would have required difficult, even controversial, proofs.

Set theory was axiomatized in 1908. Through the vicissitudes of the language, which we are glossing over, the word "axiom" replaces the more ancient "postulate"; an axiomatized theory is a theory organized like one of Euclid's, with the deduction of propositions from a system of axioms. However, not only at the time of its origins but even after it was axiomatized, the proofs of set theory have not been rigidly structured like those of Euclid's *Elements*. Many paths cross, intersecting like a net, and it's possible to trace various routes that all lead to the more important, unavoidable nodes. For every story of a proof, there are always others with different premises and assumptions.

Cantor   There is no one-to-one correspondence between real numbers and natural numbers.

The first proof was reductio ad absurdum. The version that we summarize, the published version, is much simpler than the one he included in his letter to Dedekind.

Cantor assumes that there is an enumeration $\theta_1, \theta_2, \ldots, \theta_n, \ldots$ of all real numbers. He intends to prove that within each interval (α, β), with $\alpha < \beta$ there exists a number η that is not part of the enumeration.

Given (α, β), Cantor considers the first two elements of the enumeration (i.e., with the least indices) that fall into this range, θ_{n_1} and θ_{n_2}, and with these extremes identifies an interval (α_1, β_1) contained in the previous one. Then he repeats to obtain (α_2, β_2), (α_3, β_3), etc.

If the production of intervals ends with an (α_n, β_n), this means that at most one element of the enumeration can fall within this interval; therefore, due to the density of numbers, you can find there a number η not belonging to the enumeration. On the other hand, if the procedure continues infinitely, the leftmost ends of the intervals must have a least upper bound α_∞ and the rightmost ends a greatest lower bound β_∞, with $\alpha_\infty \le \beta_\infty$.

If $\alpha_\infty < \beta_\infty$, the reasoning is subtle; it concerns the indices. Suppose a θ_{n_0} is in all (α_n, β_n), so in particular it is neither an α_{n+1} nor a β_{n+1} for every n. Therefore in forming an $(\alpha_{n+1}, \beta_{n+1})$ two θ_i are chosen with $i \ne n_0$ (i less than or greater than n_0); however, for every n_0, there are only a finite number of indices $<n_0$. So, from a certain point onward $(\alpha_{n+1}, \beta_{n+1})$ is formed by choosing two θ_i with indices greater than n_0, and this means that θ_{n_0} doesn't belong to (α_n, β_n), otherwise it would have been one of those chosen. Contradiction.

If $\alpha_\infty = \beta_\infty$, this number cannot be part of the enumeration, and it is the η we are looking for. In fact, on the one hand it must fall within all (α_n, β_n); on the other, with the same reasoning as before, if it were a θ_{n_0} then for sufficiently large n all of the (α_n, β_n) would be formed with elements of the enumeration with an index larger than n_0, and therefore θ_{n_0} would be excluded from all of the sufficiently large (α_n, β_n) and would be external to them. □

As we can see, the density of $\mathbb{R}$ is used in an essential way, that is, the property that says that other points will fall between any two given points, and also continuity, in the form of the existence of a least upper bound for each sequence of points with an upper bound. What isn't seen is any set-theoretic symbol or concept: an enumeration or a series are not yet functions of the domain $\mathbb{N}$, a segment is not a set, the relationship between (α_1, β_1) and (α, β) is not inclusion, it is expressed by $\alpha < \alpha_1 < \beta_1 < \beta$; they are concepts of mathematical analysis, undefined, or defined through synonyms, or with figures, and with an operational meaning shared by every user.

The second proof was read in 1891 at the first assembly of the Deutsche Mathematiker-Vereinigung, the mathematical association of which Cantor had been a promoter and the first president. Set theory has now made important progress, the sets are no longer called *Mannigfaltigkeiten*, or manifolds, but *Menge*. Not only sets of numbers but abstract sets are studied. Infinite ordinal numbers have been introduced and with these the cardinal numbers. The hierarchy of infinite cardinality $\aleph_0, \aleph_1, \aleph_2, \ldots$ is unlimited, but still defined with the ordinals. $\aleph_1$ is exemplified only by the second class of the ordinals, the class of countable ordinals, that is, those with a cardinality of $\aleph_0$ like $\mathbb{N}$. It is shown that the set of countable ordinals has a minimum cardinality greater than the countable. Similarly in general the $(n+1)$th $\aleph$ from the nth class of ordinals.

The new proof, published in 1892, doesn't depend on the continuity property of the real numbers, it's consistent with Cantor's increasingly abstract approach, and the proposition is in fact formulated in general not only for the real numbers but for the set of functions from a set X (also infinite) in a set with two elements: the set of functions from X in a set with two elements has greater cardinality than that of X. This guarantees that the hierarchy of the infinite cardinality is unlimited, independent of the concept of order.

Today Cantor's theorem is formulated for $\mathcal{P}(X)$, the "power" set of X, or set of all subsets of X, but at the time this concept was not yet clear (although Cantor in his letter to Hilbert of October 10, 1898 acknowledges that the totality of the subsets of the natural numbers is cardinally equivalent to the totality of the functions in $\{0, 1\}$). The formulation of the theorem in terms of $\mathcal{P}(X)$ was independently carried out by Ernst Zermelo and by Bertrand Russell (1872–1970).

Cantor The set of functions from X in a set with two elements has greater cardinality than that of X.

The proof is as follows: Given a set L and the set M of the functions from L in $\{0, 1\}$, it is obvious that the cardinality of L is less than or equal to that of M, because the functions that have the value 1 for a single element of L form a subset of M equipotent to L. It is then sufficient to demonstrate that the cardinality of L and M are different to conclude that the power of M is strictly greater than that of L. At that time, the comparability of any two sets with respect to cardinality, called trichotomy, was admitted, even if it had not yet been proved, and indeed would have required a very demanding proof.

It is assumed that the cardinalities of L and M are equal, therefore that there is a bijection between the two sets, and the function corresponding to the element $l \in L$ is indicated by f_l. Now we build a $g : L \to \{0, 1\}$ different from all f_i, putting

$$g(l) = \begin{cases} 1 \ \textit{if } f_l(l) = 0 \\ 0 \ \textit{if } f_l(l) = 1 \end{cases}$$

where so-called "diagonalization" is recognized. In the case of real numbers, assuming that there exists an enumeration $\{r_n\}$ of the real numbers in $(0, 1)$, given for example in the form of infinite sequences of 0 and 1, each a function from $\mathbb{N}$ in 2, we define a sequence not belonging to the enumeration by changing the "diagonal" of the infinite matrix with the exchange between 0 and 1.

It's easy to check that a contradiction occurs: g must be a f_{l_0} and $g(l_0) = 0$ if and only if $g(l_0) = 1$.

It would be enough to assume, by reductio ad absurdum, that there is an injection of M in L, that is to say that the power of M is less than or equal to that of L. Given such an injection f, to each element l of *the image* of f there is a corresponding function $f_l = f^{-1}(l)$. Define g as above (at least for l in the image of f), and otherwise arbitrarily, obtaining an extension $\bar{g}$. But $f(\bar{g}) = l_0$ and $\bar{g} = f^{-1}(l_0) = f_{l_0}$, and $\bar{g}(l_0) = g(l_0) = f_{l_0}(l_0)$, from which $g(l_0) = 1$ if and only if $f_{l_0}(l_0) = g(l_0) = 0$.   ⊔

Apart from its mathematical value, Cantor's theorem played an important role in the history of the foundations of mathematics. In 1901, Bertrand Russell had discovered (as had Zermelo, perhaps before him) the paradox of the class of all cardinals on his own (Russell used the term "class" rather than "set"). Initially, Russell believed that it showed an error in Cantor's theorem; he was convinced that the universal class was legitimate and therefore a greatest cardinal existed, while the theorem implies that for each cardinal there is a greater one. Russell (1901) speaks of Cantor's "subtle error." When the article was reprinted (in Russell 1918), he added some notes written in 1917: in one of these Russell corrects himself about Cantor's error, stating however that the explanation would be too complex to show. In looking for a mistake in Cantor's proof, which he reformulated to apply it to $\mathcal{P}(x)$ instead of to functions, he was led to consider the class of classes that do not belong to themselves.

In fact, Cantor's proof that we've seen above, adapted to $\mathcal{P}(x)$, leads to these considerations. Assuming that there exists an injection f from $\mathcal{P}(x)$ into x, we arrive at a contradiction by defining a subset z that is different from all subsets of x. Now each subset is $f^{-1}(y)$ for some $y \in x$, and because the subset sought is different from $f^{-1}(y)$ you can play on y, putting it in z if it is not in $f^{-1}(y)$ and not putting it in z if it is in $f^{-1}(y)$; it is therefore defined as

$$z = \{ y \in x \mid y \notin f^{-1}(y) \},$$

which can be considered a sort of antidiagonal set. As a subset of x, z has an image $f(z)$, and $z = f^{-1}(f(z))$; but $f(z) \in z$ if and only if $f(z) \notin f^{-1}(f(z)) = z$, a contradiction.   □

In the case where x is the universal class, $\mathcal{P}(x)$ can be thought of as injectable in x with the identity function, the function i such that $i(x) = x$, and the z of the proof becomes $z = \{y \mid y \notin y\}$, which is the class z that appears in Russell's antinomy:

$$z \in z \leftrightarrow z \notin z.$$

Indeed Russell (1906) presents what he calls a "simplification of Cantor's proof" in a similar way to what is shown here. In *The Principles* (Russell 1903) §100 he also states:

> I was led to it [the puzzle] in the endeavor to reconcile Cantor's proof that there can be no greatest cardinal number with the very plausible supposition that the class of all terms . . . has necessarily the greatest possible number of members.

We can therefore say that the origin of Russell's antinomy is in Cantor, or at least in Russell's reflection on Cantor and in his modification of Cantor's proof. After accepting Cantor's proof, Russell also published the antinomy of the greatest cardinal in the *Principles*, §344ff.

Finally, we consider Cantor's other result from the 1870s:

Cantor $\mathbb{R}$ and $\mathbb{R}^n$ are equipotent for each n.

Cantor actually considered numbers and n-tuples of numbers, but we'll show the simpler case $n = 2$ and the equipotency of side and square.

In January of 1874, Cantor posed a problem to Dedekind:

> Can a surface (say a square including its boundary) be one-to-one correlated to a line (say a straight line including its endpoints) so that to every point of the surface there corresponds a point of the line, and conversely to every point of the line there corresponds a point of the surface? (Ewald 1996, 2:850)

He thought not, because at the time he believed that were it so, difficulties related to the concept of dimension would have arisen, but he was convinced that it would be difficult to find an answer. The proof of the theorem completed, Dedekind had to explain to Cantor that the result did not endanger the concept of dimension, in that this had to remain invariant with respect to continuous correspondences in both directions.

In June of 1877, Cantor told Dedekind that he had solved the problem he'd put to him, in an unexpectedly positive direction. In responding, Dedekind objected to the proof: Cantor considered two numbers from the interval $[0, 1]$ in decimal representation

$$0.\alpha_1\alpha_2 \ldots \text{ and } 0.\beta_1\beta_2 \ldots$$

and associated with them the number

$$0.\alpha_1\beta_1\alpha_2\beta_2. \ldots$$

But the decimal representation is not unique: some rational numbers have all 0s from a certain point onward,

$$0.\alpha_1 \ldots \alpha_n 000 \ldots$$

or all 9s,

$$0.\alpha_1 \ldots (\alpha_n - 1)999 \ldots;$$

if you choose a particular representation, for example excluding the sequences equal to 0 from a certain point onward, some points of the segment, those of the form

$$0.\alpha_1\beta_1\alpha_2\beta_20\beta_30\beta_40\beta_50 \ldots,$$

are not correlated to any pair. Similarly if you exclude sequences equal to 9 from a certain point onward.

Two days later, on June 25, Cantor was able to send Dedekind a new proof, with an ingenious distinction in two steps, while still remarking, "I cannot repress a certain regret that the subject demands more complicated treatment."

The new proof repeats the original in the first step, but restricted to points of irrational coordinates and using the development in continued fractions, which is unique for the irrational numbers.

The second step establishes a correspondence between the interval $[0, 1]$ and the set of irrational numbers contained in the interval $[0, 1]$, which we indicate with $[0, 1]^*$.

The proof of this is not direct. If $\{r_v\}$ is an enumeration of all the rational numbers of the interval, and $\{\varepsilon_v\}$ is any increasing sequence of irrational numbers tending to 1 $\left(\text{for example} \left\{ 1 - \dfrac{\sqrt{2}}{2^v} \right\} \right)$, you can easily establish a one-to-one correspondence between $[0, 1]^* = [0, 1]/\{r_v\}$ and $[0, 1]/\{\varepsilon_v\}$. So, if a one-to-one correspondence between $[0, 1]/\{\varepsilon_v\}$ and $[0, 1]$ is established, the proof is done.

Since $\{\varepsilon_v\}$ divides $[0, 1]$ into infinite small intervals,

Cantor says he is sure that Dedekind will have no difficulty seeing that it is sufficient to show in general that there is a one-to-one correspondence between any closed interval $[a, b]$ and the half-closed interval $(a, b]$.

Dedekind had no difficulty seeing it, but for the rest of us mortals here is a possible reduction from one problem to another. Suppose that there is a one-to-one correspondence between any closed interval $[a, b]$ and the half closed interval $(a, b]$. The correspondence between $[0, 1]$ and $[0, 1]/\{\varepsilon_v\}$ is constructed in sections, starting with $[0, \varepsilon_1)$ and sending the numbers less than ε_1 to themselves; then, taking advantage of the supposed existence of a one-to-one correspondence f_1 between $[\varepsilon_1, \varepsilon_2]$ and $(\varepsilon_1, \varepsilon_2]$, we send the elements of $[\varepsilon_1, \varepsilon_2]$ where they are sent by f_1, and a one-to-one correspondence is obtained between $[0, \varepsilon_2]$ and $[0, \varepsilon_2]/\{\varepsilon_1\}$.

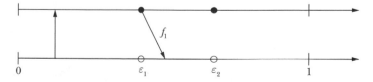

However, in order to extend the correspondence to the elements of the interval $(\varepsilon_2, \varepsilon_3]$, in the supposed correspondence f_1, the right

endpoint ε_2 must be sent to itself, as shown in the following figure. Cantor doesn't point this out, but the condition is satisfied by the subsequent proof, which we'll see later, of the existence of a one-to-one correspondence between closed and half-closed intervals. Details are explained in the long version of the story which are omitted in the shortened version and replaced by magic interventions.

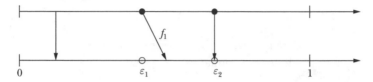

If this is so, in fact, extension is possible. When we consider an f_2 that establishes a one-to-one correspondence between $[\varepsilon_2, \varepsilon_3]$ and $(\varepsilon_2, \varepsilon_3]$, f_2 will send ε_2 into $(\varepsilon_2, \varepsilon_3)$; if we join it to the application constructed before, we would have two values for ε_2:

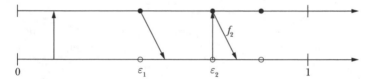

But it's enough to remove $f_1(\varepsilon_2)$, or eliminate the pair $\langle \varepsilon_2, f_1(\varepsilon_2) \rangle$ from f_1, to have from the union of f_1 and f_2 a function, and exactly a function that doesn't have ε_2 in its image, and that sends $[0, \varepsilon_3]$ into $[0, \varepsilon_3] / \{\varepsilon_1, \varepsilon_2\}$.

The correspondence between $[0, 1]$ and $[0, 1]/\{\varepsilon_v\}_v \in \mathbb{N}$ is therefore defined by iterating the procedure, after fixing for each v a one-to-one correspondence f_v between $[\varepsilon_v, \varepsilon_{v+1}]$ and $(\varepsilon_v, \varepsilon_{v+1}]$ with the property $f_v(\varepsilon_{v+1}) = \varepsilon_{v+1}$.   □

The figure below is Cantor's original, with $c = 1$ if we want to refer to $(0, 1]$ on the ordinate axis and $[0, 1]$ on the abscissa axis. To establish a correspondence between the xs and the ys we begin by establishing a linear correspondence between arbitrary $[0, b]$ and $(0, a)$ (in the figure $a = b$) by means of the oblique segment from $(0, a)$ to $(b, 0)$ and in addition associating a to 0. Then we extend it by using the segment from (b, a') to (b_1, b') with b sent to a', and iterate.

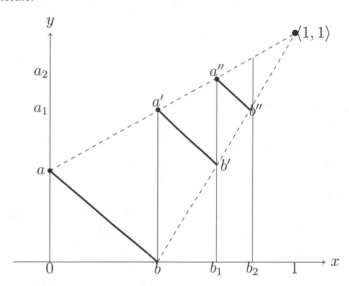

Points b, b_1, b_2 and $a, a', a'', \ldots$ are obtained by gradually halving the intervals. It is easy to represent this function analytically, but Cantor relies on the figure, eloquent in itself.   □

The ingenuity of the construction is admirable, but the overall proof was anything but quick and straightforward. Cantor himself is uncertain whether he has given a well-constructed proof. In his letter of June 29, 1877, in which he urges Dedekind to give a definitive evaluation, Cantor famously writes: *Je le vois, mais je ne le crois pas* (I see it, but I don't believe it). This confession is usually interpreted as referring to the result itself, to its paradoxical character. But Cantor, while aware of the unexpected nature of the theorem, doesn't write this phrase when he presents the result, but rather when he is clearly worried about the proof, which is not convincing beyond all reasonable doubt:

> the communications which I lately sent you are even for me so unexpected, so new, that I can have no peace of mind until I obtain from you . . . a decision about their correctness. So long as you have not agreed with me, I can only say: *je le vois, mais je ne le crois pas.* (Cantor and Dedekind 1937; English translation in Ewald 1996, 2:860)

It makes more sense to read this phrase in the context of the proof. In fact, the result is anything but evident; one can see the individual steps of a proof but not find them convincing enough to believe in the result, without the endorsement of an expert in whom one has confidence. Dedekind replies that he didn't find any gaps in the proof, that he is convinced that the interesting result is correct, and congratulates Cantor.

There were not yet the general theorems regarding sets that allow us to say, for example, that the difference between a set of a given infinite cardinality and one of a lesser cardinality has the same cardinality as the first, or more simply that adding an element to an infinite set doesn't change its cardinality, which would be useful for the lemma on intervals. Cantor had to use tools that weren't suited to the nature of the problems he was facing. The theorems would come from the theory that Cantor had not yet begun to construct. The hope that Cantor expressed, saying (in his letter of June 25, 1877) "perhaps it will later turn out that the missing portion of that proof can be settled more simply than is at present in my power" (Ewald 1996, 2:856), will then be fulfilled.

Already in October of 1877 Cantor had found a simpler proof, one that was easier to generalize in set-theoretic terms. Cantor was aware that $\mathbb{N}$ can be broken down into two disjoint subsets (the even and the odd) and takes advantage of this property in the new proof; he always uses the language of analysis, that is, speaks of a variable e which takes all the irrational values >0 and <1, and of an

x which assumes all the values in [0, 1], and writes $e \sim x$ to indicate that the values of the variable e can be correlated one-to-one with those of the variable x. The proof then reads like this:

Let φ_v be the sequence of rational numbers in [0, 1] and η_v any sequence of irrational numbers; let h be a variable that takes all values in [0, 1] except those of φ_v and η_v. Then [with his particular notation for what we would call the union] x is the union of h, φ_v, and η_v while e is the union of h and η_v. The latter can be considered the union of h, η_{2v}, and η_{2v+1}, and now it's obvious how to establish the one-to-one correspondence: h is sent to h, φ_v to η_{2v}, and η_v to η_{2v+1}. □

Far from being flawed, the original demonstration is beautiful, pulled off in a long, breathless run. But, as he predicted, this theorem of Cantor's is now relegated to the exercise book. How would a student solve this exercise? Both the lemma and the equivalence between [0, 1] and [0, 1]* depend on another observation: that every infinite set contains a countable subset. If we admit this, the proof that if x is infinite and $a \in x$ then $x/\{a\}$ is equipotent to x is reduced to the proof that $\mathbb{N} \cup \{a\}$ is equipotent to $\mathbb{N}$: if $a \in \mathbb{N}$ it is obvious, if $a \notin \mathbb{N}$ just send a into 0, 0 into 1, 1 into 2, and so on. So, by directly calling $\mathbb{N}$ a countable subset of $x/\{a\}$,

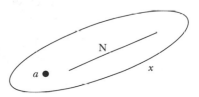

we have that $\mathbb{N} \cup \{a\}$ is equipotent to $\mathbb{N}$, and $x/(\mathbb{N} \cup \{a\})$ is equipotent to itself with the identical function, so x is equipotent to $\mathbb{N} \cup (x/(\mathbb{N} \cup \{a\})) = x/\{a\}$. □

Our student has edited out a part of the story that perhaps disturbs her, but it's possible that this very editing disturbs the reader. The student has performed an illusionist's sleight of hand by bringing ℕ into the scene, and since everyone is familiar with it, nobody wonders where it comes from.

Every infinite set x, which as such (by definition) has an injection θ in itself that is not surjective, with an element a not belonging to the image of θ, allows for the definition of natural numbers to be replicated within it: the intersection of all subsets of x that contain a and are closed with respect to θ. Let's call it $\mathbb{N}_x$. $\mathbb{N}_x$ is isomorphic to ℕ by means of a bijection. But where is ℕ? ℕ is nowhere: what we call ℕ is the $\mathbb{N}_z$ of a set z, the existence of which is affirmed by the axiom of infinity. Since all $\mathbb{N}_x$ are isomorphic to each other, any one of them can be called ℕ, and is called ℕ, and ℕ is any of these. ℕ, which is so unique, so individual and fundamental, it too "propagates" into many replicas, multiplying itself for every infinite x. It's fantastic, or horrible. Euclid had never dealt with such dramatic ambiguities: he only had some difficulty in defining the equality of angles without overlapping them; yet he also calls polygons that are equivalent "equal."

We say that ℕ is unique up to isomorphisms. And, note, even 0 doesn't exist, it's not true that it's ∅, 0 is the 0_x of any $\mathbb{N}_x$, infinite x.

Can we fix an $\mathbb{N}_x$ by convention? Yes, we usually choose an infinite x that contains ∅ as an element and is closed with respect to the function $\alpha \to \alpha \cup \{\alpha\}$ (a set of ordinals). Then 0_x becomes ∅. But when the student resolves the proposed exercise, instead of ℕ and a she should talk about $\mathbb{N}_x$ and transport the whole arithmetic structure from the set where she studied it to $\mathbb{N}_x$ by means of isomorphism. The notations and the expressions make the discourse cumbersome, so the student is granted the linguistic license she has used.

The example illustrates how mathematicians are able to shape a language to talk about that which it seems impossible to talk about; but they're also capable of creating new rhetorical forms (such as: unique up to isomorphisms, which would be like saying, or would imply: unique but also not) to overcome the language's apparent inconsistencies. Even the points at infinity of projective geometry, mentioned in passing in section 3.3, require an expedient linguistic invention (homogeneous coordinates).

That Cantor's proof becomes an exercise is a typical case of how proofs are simplified with the enrichment of a theory. However, the opposite also happens: the proof starts out simple but mathematicians look for others that establish links with advanced concepts, as we said in the introduction. A short example is Euclid's theorem on the infinity of primes: Prime numbers are more than any assigned multitude of prime numbers (*Elements* IX.20).

The proof: given a finite number of prime numbers, $p_0, p_1, \ldots, p_n$ (Euclid considers $n = 3$ to illustrate the reasoning), take the product of these and add $1: p_0 \cdot p_1 \cdot \cdots \cdot p_n + 1$. This is a number that is not divisible by any of the numbers in the sequence $p_0, p_1, \ldots, p_n$, otherwise that divisor would also be a divisor of 1; but the number is either prime or has a prime divisor (*Elements* VII.32), and therefore there are primes other than those given (we'll also talk about this in the conclusions at the end of this book). □

In 1737 Leonhard Euler (1707–1783) proved the formula for the function of the complex variable ζ:

$$\zeta(s) = \sum_1^\infty \frac{1}{n^s} = \prod_p \frac{1}{1 - p^{-s}} \qquad [1]$$

where p varies on the prime numbers; from [1], for $s = 1$,

$$\sum_1^\infty \frac{1}{n} = \prod_p \frac{1}{1 - p^{-1}}. \qquad [2]$$

The series on the left is the harmonic, divergent series, so the product on the right must also be divergent, which wouldn't happen if prime numbers were finite.

Once you have the idea, the direct demonstration of [2] isn't difficult and you can bring it back to elementary knowledge. The harmonic series for the prime factorization theorem can be considered the product of the series

$$1 + \frac{1}{p} + \frac{1}{p^2} + \frac{1}{p^3} + \cdots$$

for the p primes, each geometric series equal to $\dfrac{1}{1-p^{-1}}$.   □

Five other proofs of the theorem are found in Aigner and Ziegler (2006).

2.2   Metaphors and images

Simon Stevin (1548–1620) introduced his 1583 book on geometric problems (*Problematum geometricorum*) with a poem by Lucas Bellerus, which describes mathematics as the art of personally exploring the dark, hidden recesses of nature, revealing its secrets:

> *Vere igitur Diuam veteres dixere Mathesin Cuius ab arte labor,*
> *superas cognoscere sedes Terrarum, pelagique vias, et aperta tene-*
> *bris Natura secreta dedit.*

("Truly, then, the ancients called / Divine mathesis that which by / Its craft enabled to recognize / The supreme seat, the ways of the earth and sea / And to see in person the hidden places in the dark / The secrets of nature.") Bellerus's poetry is cited by Amir Alexander (1963–) as the work of "Luca Belleri" and can be read in Stevin (1958, 144). Lucas Bellerus belonged to a family of printers from Antwerp, where the book was published by his brother Joannes Bellerus. The "Lucae Belleri Carmen" on the frontispiece is probably an homage by the printers.

Elsewhere Stevin compares himself to a sailor who found an unknown island. The model is that of the explorer, the imperative that of direct experience, according to a philosophy of knowledge supported by Francis Bacon (1561–1626) and Galileo Galilei (1564–1642) (Alexander 2012, 11). Stevin is not an isolated case. In the letters between Galileo, Bonaventura Cavalieri (1598–1647), and

Evangelista Torricelli (1608–1647) the theme of travel and exploration returns. Cavalieri quoting Horace (*Carmina*, ode III) recognizes that Galileo was "the first to dare to steer the immensity of the sea and plunge into the ocean" (Galilei 1890–1909, 18:67, letter 3889). Torricelli congratulates Cavalieri for having opened the "royal road" to mathematics with his method of indivisibles, which brings the mathematical explorer to the heart of impenetrable problems with direct, brief, and positive proofs, and helps him to trace the map of his terrain full of wonders, while risking shipwrecks and paradoxes (Torricelli 1919, 382).

"The story of the mathematician as a maritime explorer is a vivid example of a mathematical tale that draws on historical circumstances" such as famous voyages of geographical discovery, favored by the fact that many mathematicians were actively involved in these voyages, providing navigational instruments and maps or even in person, like Thomas Harriot (1560–1621) in England (Alexander 2012, 17).

In a completely different way, in the early nineteenth century, stories such as those of Évariste Galois, who died in a mysterious duel, or Niels Henrick Abel (1802–1829) who died of tuberculosis—both with their works buried under the mountain of papers at the Académie, lost or not understood—offer the romantic model of the tormented, unhappy and unfortunate hero who is still capable of reinvigorating mathematics. Abel and Galois didn't study the secrets of nature, they planted the seeds of new, future mathematics, with a painful visionary passion. Augustin-Louis Cauchy (1789–1857) and Siméon-Denis Poisson (1781–1840) are the distracted academics responsible for the foundering of Abel and Galois's works, and could, in their distraction, be taken symbolically as an emblem of refusal of the new.

We can add Farkas Bolyai (1775–1856) who in 1829 admonished his son János (1802–1860) to give up on the fifth postulate, taking up the metaphor of the journey, now immersed in the livid atmosphere appropriate to the times: "I know the way until its end—I also have measured this bottomless night. I have lost in it every light, every joy of my life. . . . You should shy away from it" (cited from Struyk 1981; Postulate 5 states that if a straight line falling on two straight lines makes the interior angles on the same side less than two right angles, the two straight lines meet on that side). As his father feared, János was not the first to find new geometries. He was preceded by Nikolai I. Lobachevsky (1792–1856) a year earlier in a Russian publication and humiliated by Carl F. Gauss (1777–1865) who claimed to have already done everything the younger man had and, with something like pity, acknowledged only that János had saved him the effort of writing. On the other hand, in this case the dark road, as far as Farkas is concerned, is a romantic formative path rather than a path of discovery.

From these examples it is easy to recognize that, more than just mathematical tales, those mentioned are metaphors with which mathematics and its devotees are represented in certain periods, images based on the ideas and visions that most affected the imagination at that time. When it comes to the eighteenth century, Alexander dwells on the figure of Jean-Baptiste Le Rond d'Alembert (1717–1783), described at the time as a natural man, pure-hearted, not taken up with society's rivalries and jealousies. The eighteenth-century mathematician is a person in tune with the world's natural rhythms and harmonies.

Of course metaphors can cross the boundaries of historical periods. We've seen how Bolyai reprises that of the journey—of the unmapped road where you must extricate yourself from impediments and obstacles—which is one of the most common. Maryam Mirzakhani (1977–2017), the first woman to have won the Fields Medal, admitted that when she had a result she felt as if she were on the top of a hill from which the entire plain could be contemplated and enjoyed; however, when she was in the midst of research the feeling was rather like one of "being on a long hike with no trail and no end in sight" (Mirzakhani 2014). The metaphor of the sea voyage ended up a bit watered down in its dissemination, as Eugenia Cheng (1976–) affirms:

> I love abstract mathematics in somewhat the same way that I love boat trips. It's not just about getting to a destination. . . . It's about the fun, the mental exertion, communing with mathematical nature and seeing the mathematical sights. (Cheng 2017)

Although they capture a mood or an attitude, metaphors in themselves don't develop into a story; but, as Doxiadis and Mazur affirm in commenting on Alexander (2012):

> Stories play crucial roles in our discovering, creating, explaining, and organizing knowledge, and thus mathematics also has a great need for narrative, even though its taste for general ideas might make one forget this. (Doxiadis and Mazur 2012, vii)

As we'll see in section 2.4, for Aristotle, on the contrary, it is precisely the universal that constitutes a link with poetry.

Apostolos Doxiadis (1953–) and Barry Mazur (1937–) edited *Circles Disturbed* (2012), a rich collection of essays showing, through a variety of perspectives, an explicit conviction regarding the close relationships between mathematics and literature. The book's title refers to the last words Archimedes is purported to have said to a Roman soldier before he was killed: "Don't disturb my circles." Later on we will discuss other contributions in this volume, in addition to Alexander's. Doxiadis is also the author of, among other things, the successful novel *Uncle Petros and Goldbach's Conjecture* (2000). Mazur is a specialist in Diophantine geometry and author of the brilliant *Imagining Numbers* (2003).

The authors who contributed to the volume don't always manage to adequately justify their vision of the relationship between mathematics and stories. For example, Alexander hints at the possibility that the transformations which occurred in mathematics between the seventeenth and nineteenth centuries were partly guided by the form of the stories that accompanied them. However, Alexander himself must honestly admit that the decisive difference between his mathematical explorers and their successors is that the former didn't have tried-and-true mathematical tools, and had to construct them amidst difficulties and uncertainties. The tools are not produced by the stories but vice versa. What Doxiadis and Mazur say is more seductive. According to them the stories, be they of exploratory voyages or tormented destinies, are not only the retelling of events that occurred but "play a crucial role in forming the meaning of such events."

2.3   The meaning of stories

So the meaning of events would be shaped by the stories that relate them. In order to explore the validity of this thesis, we must broaden its scope to each sequence of events and each verbal description of them. We must incorporate a reflection about stories of any kind, including historical accounts, scientific reports, chronicles; in essence, all discursive activity. The digression will carry us to the shores of pure philosophy.

Two quotes from the literary field will guide and reassure us. Antonio Tabucchi (1943–2012) hypothesized that literature gives meaning to life; to make sense of the disparate sequence of events that occur we must cover them with words, "that is, 'tell the story of life.' In telling it, a meaning is obtained. But what is the ultimate meaning of this story? It is the story itself" (Tabucchi 2013).

Speaking about the writer Javier Marías (1951–), Gian Luigi Beccaria (1936–) observes:

> The writer knows that life and its stories, without the filter of literature—which retells, articulates, and reorders them—are, for both the author and the reader, *per se inert materials*, without tension, because they lack the rhythms and flow of images in which the writer knows how to immerse you. It is necessary for the *random and unresolved flow of life to be filled with meaning*, because stories almost always seem meaningless, and perhaps they are, but the writer

is always there to show us the strength of narration. (Beccaria 2016, 150–151, italics mine)

With the aim of understanding what Tabucchi and Beccaria mean, let's do a thought experiment. Let's go directly to the nonartistic level of common discourse and ask ourselves how we would manage if we had only a physicalist language, like the one Rudolf Carnap (1891–1970) wanted to construct (see appendix, 2). We should then describe, for example, how the fingers of one hand bend inward as far as possible toward the palm and come together while the arm to which they belong quickly extends until it forcefully meets the face of a person, or something like that, in order to say that someone has been punched. But for us, "to punch someone" has a meaning full of information that doesn't concern the muscles but rather the feelings involved in the act; emotions that describe a reality that can't be reduced to the physical: let's say a moment in a violent confrontation, or a quarrel. Feelings are real, they guide us and drive us to action; but, unless we personify a myriad of spiritual entities, we must say, as Tabucchi does, that their reality exists only in the story of their manifestation.

In confirmation, let us reflect on the relationship between languages for specialized scientific purposes and for communication. Consider the languages used in computer science and neurophysiology (the following is a summary of some points in a discussion between Riccardo Manzotti and Tim Parks in Manzotti and Parks 2016–2017). In describing what a brain does, just as in describing what a computer does, everything can be presented comprehensively in terms of causal processes, chemical reactions, and changes in voltage. If a neuroscientist observes what happens in a mouse's brain when it's near a piece of cheese, complex chemical and electrical activity can be detected; perhaps the results are also accompanied by some magnetic resonance images; however, the conclusions are presented using the language of information processing. The researcher will tell us that the entorhinal lateral cortex calculates and transfers information from the olfactory bulb to the hippocampus and then other brain neurons are activated to produce salivation. What physicalist language provides us with is only a precise description of the activity of a few neurons; however, thanks to the narrative of the scientist who used the word "information," we perceive a familiar situation even if we don't know exactly where the entorhinal cortex and the hippocampus are. We understand what is happening: the mouse smells the cheese—or food, we don't know if the word "cheese" appears on the screen of its mind, as Christopher John Francis Boone describes in *The Curious Incident of the Dog in the Nighttime* (Haddon 2003)—better yet: it smells, and that awakens the desire for food and its mouth waters. Or let us say that we believe we understand, because for us "information" is always information about something, the transmission of understandable content that increases our knowledge. From other similar cases we understand more generally that the brain stores memories, decodes representations, processes information, and produces consciousness.

We focus on the concept of information because it's a term that has a scientific theory, as well as being an everyday word. Information is technically the ability of a channel to establish a causal link between two events, in source and reception, speaking and listening, writing and reading (in the metaphorical sense of Turing's machines: writing is only producing a sign, reading is reacting differently to different signs). There is nothing in between; if there is no reception upon arrival, the information doesn't exist. It's not like it leaves the source and then gets lost somewhere along the way.

However, the set of stories related to the production, transmission, and reception of information—including that of the mouse receiving information about the proximity of food from its nose—leads us to imagine that there is something mental, rather than physical, produced by the brain. It is as if within the brain—in addition to the organic material—an immaterial "information," the content, the meaning of information, existed or emerged. From its distant Cartesian origins, with other related problems, dualism is at the center of the popular, contemporary discussion of consciousness.

When we interpret information as the transmission of meaning, we superimpose our deep-rooted dualism on the mathematical theory. In fact, the central dogma of information theory is "meaning is irrelevant." So much so that it is possible to transmit information accurately even using communication systems with a lot of noise; all that is needed is an adequate amount of redundancy: the message must be repeated a sufficient number of times. All that is necessary for the transmission of information is coding, or two systems: one at the input source (encoding) and one at the output source (decoding); the encoding and decoding are syntactic translations between formal languages.

But we don't yet know the location of the translator that transforms the neuronal activity in the hippocampus into the phrase "cheese smell," or rather, into the activity of other neurons devolved to language which form the phrase "cheese smell" in the alphabet of the mouse's mental language; nor do we know about those that receive (understand?) the phrase and activate the response. In the case of humans, there is probably more than one translator; a hierarchy from "mentalese" to natural language. Therefore we don't know who reads it or listens to it, at the final level, and understands it: let's say our consciousness, or the mouse's equivalent. We spontaneously imagine a *homunculus* (or a *musunculus*, from mouse: *mus*, *muris* in Latin) that understands the phrase. It is almost certainly the (human) brain that, knowing how to do the mathematics necessary for Gödelization, builds a self-reference—as with the phrase "this sentence is . . . ," "I am. . . ." Self-reference isn't dangerous. In the case of "I am a liar" you simply get a statement that is neither true nor false. But what do neurons know about truth and lies? Gödelization doesn't use any foreign concepts, only elements of the given language. Self-reference only closes the cycle of syntactic translations from one language to another, which would induce an infinite continuation; a phrase that refers to itself makes sense, it is no longer just a string of symbols.

Therefore, superimposed on observation, the story allows us to understand but, at the same time, suggests tacit adhesion to a dualistic philosophy of mind which, analyzed by specialists, has many weak and controversial points. It happens like Platonism in mathematics. At first it seems natural enough (the existence of numbers, functions, spaces) but, if critically analyzed, is full of flaws and naive assumptions. Furthermore, the scientific account becomes tinged with teleological accents, rather than being limited to purely causal relationships. The distinction between causal and teleological explanations goes back to Aristotle and is still vigorously argued, not only in the philosophy of mind; a similar situation occurs with regard to presentations of Darwinian evolution. The teleological explanation (that which uses the concept of purpose in answering the questions "why? how come?") comes naturally when we're talking about something like an organ and its adaptation to a different function. These expressive forms sometimes seem to imply that the end result of the adaptation is due to intelligent design.

Richard Dawkins (1941–), champion of evolutionary theory, admitted that "the illusion of design conjured by Darwinian natural selection is so breathtakingly powerful" (Dawkins 2016, 633). However, according to Darwinian theory, functions and organs don't exist. Most biologists would say that the transposition into these words is a convenient way of expressing themselves but that functions and purposes don't exist in nature. The scientific version should eliminate them with a reduction to physicalist language. On the other hand, Daniel Dennett (1942–) protests that there is design, and we shouldn't be ashamed to use this word nor should we renounce the use of other teleological terms. What needs to be done is to explain that there is design in nature, and it would be ridiculous to deny it in the face of so much evidence, but no designer. The design is not in someone's mind but formed through the mechanism of evolution, rightly understood (see the discussion in Dennett 2017, 32–43).

With our thought experiment, perhaps we've made plausible Tabucchi and Beccaria's thesis, about the need to use narration to give meaning to the random, mute flow of life.

Through several of its representatives the literary world has certainly spoken out in favor of their vision. This theme is usually dealt with in the context of discussion about the relationship between biography, or autobiography, and imagination (a topic extensively explored in chapters 4 and 5 of Lolli 2017). For example, Lalla Romano (1906–2001) sees no difference between invention and testimony:

> It is believed that fantasy means invention, while on the contrary, memory is testimony. But for a writer, memory and fantasy are the same thing. . . . I believe that everyone can only invent stories if he or she has already lived them as if they were, precisely that, invented, a fantasy. (Romano 1998, 91)

In the course of explaining why he used a female alter ego to tell the story of his escape from a concentration camp when he was five, Aharon Appelfeld (1932–) said that reality is inexplicable unless narrated:

> Reality, as you know, is always stronger than the human imagination. Not only that, reality can permit itself to be unbelievable, inexplicable, out of all proportion. The created work, to my regret, cannot permit itself all that. (Roth 2010, 74–75)

2.4 Proofs and stories

It's commonly thought that mathematics—eternal, transcontinental, irrefutable—is objective; that it describes a reality often considered more solid, more definitive than our crumbling one, indeed perhaps the only reality not subject to deterioration in substance and value, be it a parallel universe or the miraculously absolute product of our brains' eighty-six billion neurons.

Nevertheless, mathematics is stored in documents that, as we go back in time, become increasingly fragmented and incomplete, the "different arrangements of twenty characters" difficult to decipher, but yet still available and accessible in modern translations. They achieve their aim "of talking with those who are in India; of speaking to those who are not yet born and will not be born for a thousand or ten thousand years" (this is of course the famous homage to writing at the end of the first day of *Dialogue Concerning the Two Chief World Systems*, Galilei 1967). Writing makes it possible to communicate the "deepest thoughts to any other person, though distant by mighty intervals of place and time." But, in the moment that mathematics is created, it is more often spoken than written, discussed as well as thought, and it is a social product. It generally originates not from a text but from a discussion, or a set of verbal exchanges.

> When I started as a graduate student at Berkeley, I had trouble imagining how I could "prove" a new and interesting mathematical theorem. I didn't really understand what a "proof" was. By going to seminars, reading papers, and talking to other graduate students, I gradually began to catch on. Within any field, there are certain theorems and certain techniques that are generally known and generally accepted. When you write a paper, you refer to these without proof. You look at other papers in the field, and you see what facts they quote without proof, and what they cite in their bibliography. You learn from other people some idea of the proofs. Then you're free to quote the same theorem and cite the same citations. You don't necessarily have to read the full papers or books that are in your bibliography. Many of the things that are generally known are things for which there may be no known written source. As long as people in the field are comfortable that the idea works, it doesn't need to have a formal written source.
>
> At first I was highly suspicious of this process. I would doubt whether a certain idea was really established. But I found that I could ask people, and they could produce explanations and proofs, or else refer me to other people or to written sources that would give explanations and proofs. There were published theorems that were generally

known to be false, or where the proofs were generally known to be incomplete. Mathematical knowledge and understanding were embedded in the minds and in the social fabric of the community of people thinking about a particular topic. This knowledge was supported by written documents, but the written documents were not really primary. . . . I was interested in geometric areas of mathematics, where it is often pretty hard to have a document that reflects well the way people actually think. (Thurston 1994, 168–169)

William Thurston (1946–2012), the author of this passage, was a mathematician who was awarded a Fields Medal for his work in low-dimensional topology.

To keep up to date, remember that there are dedicated discussion websites like *MathOverflow* or blogs, like the Gowers's Weblog, where theorems have been demonstrated with the collective collaboration of the people who Tim Gowers (1963–) himself has asked to participate. New forms of communication and interaction are created, but the topic is secondary in this context (at the end of the chapter we'll add some considerations on mathematics as a social product).

The fact that mathematics is born and consolidated through verbal discourse isn't worrisome, because even fairy tales, myths, or epic poems, such as the works of Homer, have been transmitted orally for centuries with anonymous contributions, changes, and additions. The same oral origin sees the entry of the fantastic into medieval literature with the stories of Tristan at the beginning of the twelfth century. In this regard, Varvaro (2017) documents the appearance of similar motifs in different contexts, which are not culturally linked, testifying to an underground tradition capable of reemerging in recognizable forms. The same happened with the multiple and independent origins of mathematics on different continents.

Once written, these things stopped changing, at least once the philologists agreed on canonical versions (interpretations can still change, according to the state of criticism). Mathematics, however, is more volatile. It continues to change even after it has been written down. Every proof done by a teacher in a classroom, along the guidelines of a manual, is different from all of the others written or expounded by other teachers; it's a creative interpretation. The deliberate search for a new proof of a known theorem is a driving force behind the growth of mathematics, which doesn't find many parallels in literature. In fact, the idea is taken to the extreme by Jorge Luis Borges (1899–1986) in his "Pierre Menard, Author of the Quixote," where Menard doesn't want to "compose another *Quixote*—which is easy—but *the Quixote itself.*" New proofs are sought that are simpler, more comprehensible within the context of the new prevailing frameworks, more elegant. Whether the theorems with new proofs are still the old theorems or are now new theorems is debated. The answer depends on how the new proof places the theorem—differently, or not—in another theoretical framework, where it comes to have a function that is not the one it had in the first. A theorem is always a link between propositions that contribute to forming a network of connections, and its placement at a given point changes its meaning.

Turning to literature, we could take the example of the fables of Aesop (620–560 BCE) and their rewriting by Jean de La Fontaine (1621–1695); the interest of this work lies in the fact that we recognize the transformation of the morality associated with the rewriting. If the morality is different, the fables are different, regardless of the appearance of the same characters and situations. Furthermore, in this example we see how style, often conditioned by language, can define and determine the content (observations of Charles Rosen (1927–2012) in Rosen 2012). If the proof of a theorem is re-presented in a new language, and therefore with different assumptions, it has another content.

Of course mathematics produced in a dialogical manner must then be written down, or recorded in some way so that it can be passed on. And this must be done quickly to avoid what happened in the case of Bolyai. But what is written down, with effort and reluctance by those who put their names to the work, doesn't correspond to its construction: on the one hand the linear text is often longer, even boring, because it has to include the prerequisites and justifications for all of the inferences down to a level of elementary knowledge and arguments which is lower than that of the originators, who are all working from an adequate knowledge base. On the other hand, the writing leaves some gaps, precisely in the salient moments when decisive obstacles are encountered and overcome in the process; it is silent about the conditions that suddenly illuminated the way, moments of creativity, or random errors, the cases of serendipity, which lead research in the right direction.

For example, Cédric Villani (1973–), who recounted his proof of Landau's nonlinear damping and the convergence to equilibrium of Boltzmann's equation, recalls how a colleague, misunderstanding a diagram on Villani's blackboard, suggested a link with Kolmogoroff-Arnold-Moser's theory on the stability of the solar system, "as though Darwin had guessed correctly about the evolution of species by comparing bats and pterodactyls, mistakenly supposing that the two were closely related" (Villani 2015, 18–19). In fact, although perhaps Villani doesn't know it, an error accounts for Darwin's first shift toward an evolutionist position: he had been convinced, or erroneously assured, that *Macranchenia* was an ancestor of the current guanaco (a llama-like mammal) (Gould 1996). The temporal proximity seemed to him a completion of the spatial proximity detected in the Galápagos. Villani also expresses regret that, in the final draft of an "illumination" which allowed him to overcome a block, "technique is the only thing that matters in a proof" and that the reader will not be able to appreciate the euphoria he experienced (Villani 2015, 143). "Technique is the only thing that matters" means replacing an analogy or metaphor that produced or transposed an idea with the detailed justification of the idea from its internal presuppositions alone, or, as in the case of Villani, replacing an image whose observation provoked the idea with the analytical relationships between the elements of that image.

Anyway, dialogic activity takes place between equals, or near-equals.

Myths such as that of Socrates, in *Meno* 82–85b.10 (Plato 1977, vol. 2), in which he leads a slave to discover a proof, are in fact just myths (of some use in teaching). Socrates intervenes with the "right" diagrams and questions, that is, those suitable for recognizing the dead ends, and with the final image showing the solution. The theorem establishes that the length of one side of a square with double the area of a given square is that of the diagonal of the latter. For example, at the first answer "the double," Socrates has the slave consider the square made by doubling the side, and they easily verify that this square has area four times the given square. The new conjecture, that the side should be three feet (versus the two of the given square), reveals that the area of a square with a three-foot side would be nine feet, rather than the required eight.

Socrates now appreciates the state of confusion into which the slave has fallen, because "besides not knowing, he does not think he knows," and from a place of doubt he can begin the search for the result. "I merely ask questions, and do not teach him; and be on the watch [Meno] to see if at any point you find me teaching him or expounding to him, instead of questioning him on his opinions."

This is not what happens, Socrates doesn't limit himself to asking questions; he takes up the figure of the four squares used in the first discussion on doubling the side, and draws the diagonals; "And does this line, drawn from corner to corner, cut in two each of these spaces?"

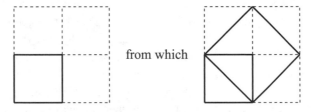

from which

The hint is too explicit. Far from proving the anamnesis theory, as he would have liked to, this maieutic experiment nevertheless confirms the dialogic nature of the proof.

What Socrates could have done, to help the slave understand the solution, would have been to draw only the diagonal

or perhaps

which would have suggested thinking of the square as composed by placing two equal right triangles side by side, with the not unthinkable inference that the double square would have to be composed of four right triangles. The problem thus becomes: can four equal right triangles be combined to form a square? Following this path, the proof could preserve the maieutic form even in a rigorous Euclidean version, where drawing (or producing) a diagram is already part of the proof. Drawing the diagonal of the given square is the winning move, and can be considered one of the premises of a proof, after the one that must come first: being able to draw a square. Nevertheless, Plato's proofs based on careful inspection of drawings until some properties are recognized are not yet those of Euclid, as is explained in Russo, Pirro, and Salciccia (2017, 14–15).

Each mathematical result is an achievement that takes place over time and through exchanges with colleagues: we think, speculate, discuss, modify, work out arguments and submit them to the attention of other researchers, or look them over again ourselves after pausing to reflect. In short, a story can be told, and sometimes you do so, at least for some of the topical moments along the way. If it's written well, the introduction of a work usually presents the description of a problem, its origins, and some stages in its development, partial solutions or results that can be related. This is done to give predecessors their due credit, or to highlight the new developments.

From this information, the genesis of the results can be partially reconstructed. That said, the authors' comments are often unwittingly misleading. For example, Kurt Gödel wrote in his thesis, omitting the observation in the printed version, that once Whitehead and Russell had decided to base logic on formal deductions from certain evident propositions, the problem of knowing whether the system of axioms and rules was complete immediately arose (see appendix, 10). In truth, eighteen years passed from the publication of *Principia Mathematica* to that of the volume of Hilbert and Ackermann where the question of completeness was explicitly posed as an open problem (Dawson 1997, 70).

Of course there are cases in which the author is isolated either because of social conditions or because of his or her character. One example is Jean-Victor Poncelet (1788–1867) who worked on his contributions to projective geometry in prison, without any interaction. We know little about Andrew Wiles, who from 1985 to 1994 was closed up in a room reflecting on the Shimura-Taniyama conjecture (the rest to get to Fermat was known). We don't know anything about the gestation of Gödel's first results, even if we can reconstruct some moments that seem important from the references he includes (or omits) in the final text; for example, the work of Skolem and that of Post and Bernays respectively. In order to understand how Gödel came to fix his interest on logic, and to pose himself the problem of the completeness of logic that he resolved in his dissertation, his biographer examined the record of which books he borrowed from the libraries in Brno and Vienna, took note of the courses and conferences he attended (those of Rudolf Carnap and Luitzen Brouwer) and the documented memories of the protagonists (Carnap's diary) (Dawson 1997, 68–69). This is a type of documentation that will no longer be available in the future, or at least not in this form.

An exception moving in the opposite direction is that of Cédric Villani, who presents not so much the proof itself but his life in the period during which it was being constructed. His attitude is:

"We owe it to ourselves to tell stories" (Villani 2015, 25), quoting Neil Gaiman (1960–), a writer, cartoonist, journalist, and television screenwriter. The result is a completely open book, without reticence, with his musical preferences, a few scenes from his family life, accounts of business trips, meetings and talks with colleagues, all interspersed with a substantial number of emails exchanged with his young collaborator, Clément Muhot (1976–), which make up a fair part of the narrative. This bears witness to the current forms of communication and scientific relationships but also, in part, to the scientific content of the exchanges that took place, which are still only understandable to experts.

However, the sense in which one could say that a demonstration is really a narration is different. Unlike what we have pointed out relative to neuroscience or evolution, the story shouldn't be superimposed on the proof, distorting it into another language. At best it could be a metalanguage that explains why and how this language is being used; for example, the metalanguage could be a language of action and the language itself a stative one: "I produce BF by extending BA" vs. "BF is an extension of BA." To Carnap's delight, the language of mathematics is bound to be formal. In neuroscience, if you don't use mentalistic concepts you end up using the language of physics and chemistry. When dealing with evolution, it takes a great deal of control to avoid teleological concepts. In mathematics, any metaphorical language can always be eliminated in favor of one formalized in the original terms of the problem; and if another language is superimposed instead (as we will illustrate in section 4.2) both are available; even a metaphorical language such as that of structures is formal.

In any case, in mathematics it's a matter of seeing how much the facts of any genetic summary to be reconstructed leave a trace in the final product as well as how extensively these traces are incorporated in what is presented as proof of the result, and therefore have the effect of making the proof understandable. In fact, one wonders whether the narrative structure is responsible for the meaning, at least in part, and whether one can ask what the meaning of proofs is. On the other hand, if you deny that they have a meaning, you embrace formalism which, in its extreme forms, is untenable and impractical. It's better to rack your brain (perhaps in vain) trying to understand what the meaning in mathematics is and where it comes from. On this occasion we limit ourselves to setting out the convincing arguments for the hypothesis that the narrative character of proofs is plausible.

The structure of proofs is commonly considered to be a deductive concatenation. But since all proofs, if formalized, use the same rules, they would then all be the same, except for the length, or the different combination—with repetitions—of the (few) rules, or the relevant axioms. The conception of the proof as an exclusive deductive link seems inadequate to cover all the functions of a proof, for example that of explanation. On the other hand, we know that in a proof we must set aside the meaning of the terms:

> During a deduction, it is legitimate and it may be useful to think about the meaning of the geometric concepts at stake; but it isn't necessary. When it becomes necessary, it's a sign of a defect in the deductions and the inadequacy of the propositions assumed to support the proof. (Pasch 1882, 82)

Pasch's declaration is the manifesto of the modern axiomatic method; see Lolli (2016a).

However, the meaning of the terms is also ignored in fairy tales: what's an ogre? It's rarely defined, apart from sometimes being described as a deformed or anthropomorphic monster; however, recurring characteristic elements are associated with it: it's frightening, ugly, bad, anthropophagous, it has superior if not magical powers, but it's also stupid, etc. We could say that it's characterized axiomatically. Then everyone imagines it as they'd like. The same is true of other fairy tale characters or elements: the king, the prince, the witch, the castle, the realm—they're never described, they are symbols, placeholders with a specific function in the economy of the story. The king must send his sons to complete a quest. The castle must contain a captive princess. In the end it's of no importance how it's built unless, being defended by a dragon, it doesn't need a large moat. Perhaps initially fairy tales had other objectives, cosmological understanding or social cohesion, but now, as Tabucchi says, the ultimate meaning of the story is the story itself. Furthermore, "narrative structures" within which they unfold "exist on their own like geometric figures or Platonic ideas or abstract archetypes and impose themselves on each single person's individual imagination" (Calvino 1995, 1964), not unlike the recurring modules in which proofs are articulated. Reversing Calvino's analogy, the geometric figures and patterns within which the proofs perform their function exist like the narrative structures of fairy tales. (For better or for worse: it seems that Francis Bacon observed that children are afraid of the dark, and fairy tales exacerbate this fear. They're also afraid of math, and it only gets worse if witches teach them.)

The analogy with fairy tales comes spontaneously for creative mathematicians. Alfred Tarski (1901–1983), who held to a nominalist materialist philosophy and was a follower of Tadeusz Kotarbiński (1886–1961), was often asked to explain the apparent contradiction between his ideas and the mathematics he cultivated, especially set theory and model theory. When questioned about these theories dealing with infinite objects—things that he did not believe existed—Tarski replied: "I believe there is value in fairy tales and in the study of fairy tales" (Burdman and Feferman 2004, 52).

Villani is also convinced that a proof can be constructed like an outlandish tale: for example, a story about galaxies, where gases and galaxies look like the same thing when you're wearing a pair of mathematical-model glasses.

Fairy tales are, in fact, a kind of spiritual guide, along with "fantastic" stories and poetry, because they are an inheritance of invented worlds—worlds parallel to or interwoven with the one in which we're living—that have no fear of incongruencies. These possible worlds manifest the way in which human beings have lived, thought, hoped, understood, or disputed with the world around them.

This is also said, at least in regard to possible worlds, by Aristotle who "distinguishes in the *Poetics* between history and poetry on the grounds that the former tells only about actuality, while the latter is concerned with all possibilities, whether they in fact happen, or fail to happen" (Smith 2016, 5). According to Aristotle the poet's task is to relate not what has happened but what may happen, and "poetry, therefore, is a more philosophical and a higher thing than history: for poetry tends to express the universal, history the particular" (*Poetics* I, 9, cited in Smith 2016, 5).

Therefore, in an ideal polarization, philosophy is certainly on the same side as poetry, even if many renowned philosophers have invited us to pay attention to that which exists individually, to the concrete. Science, at least historically, would seem to be focused on the collection of particular facts in view of a deeper understanding of the real world and would therefore be positioned on the same side as history. But contemporary science is of a different stripe; it encompasses theories that talk about dark and invisible matter, and it can be said that it creates new worlds, and even new life, through the manipulation of the existing. Mathematics would certainly be on the side of poetry because it deals with possibilities and speaks in universals. As Robert Musil decisively states: "one may call mathematics an ideal intellectual apparatus whose task and accomplishment are to anticipate in principle every possible case" (Musil 1990, 40).

He's not in bad company: Italo Calvino, wondering what might have caused the decline of interest in the fantastic in eighteenth-century Italy, replied:

> Excessive devotion to reason? On the contrary: perhaps there was too little. Fantastic literature is always—or almost always—supported by a rational design, a construction of ideas, a thought carried through to its final consequences by following its internal logic. . . . The fantastic, contrary to what one might believe, requires a clear mind,

instinctive or unconscious inspiration controlled by reason, stylistic discipline. It requires knowing how to distinguish and blend fiction and truth, playfulness and fear, fascination and detachment, at the same time; that is, to simultaneously read the world on multiple levels and in multiple languages. (Calvino 1995, 1676–1679)

Doing math means moving on different levels using different languages, using both fictions and calculations, being both poets and laborers. The "fictions" will be discussed in the next chapter.

To prevent the obvious objections to the analogy between fairy tales and mathematics, so that it can be considered judiciously—always bearing in mind the protean nature of the discipline that cannot be captured in a single scheme—we recall the reflections of John von Neumann (1903–1957) on the mathematical trends of his time. According to von Neumann, we can think that mathematics finds its origin in a relationship with the empirical world, although the genealogy is sometimes long and often obscure. "But, once they are so conceived, the subject begins to live a peculiar life of its own and is better compared to a creative one, governed by almost entirely aesthetical motivations, than to anything else and, in particular, to an empirical science." Elegance is expected from mathematics, in the architecture of theorems and theories; if the deductions are too long, a simplifying principle is expected to explain and eliminate arbitrariness:

> These criteria are clearly those of any creative art, and the existence of some underlying empirical, worldly motif in the background—often in a very remote background—overgrown by aestheticizing developments and followed into a multitude of labyrinthine variants—all this is much more akin to the atmosphere of art pure and simple than to that of the empirical sciences. (von Neumann 2004, 183)

Von Neumann ends his essay with the hope that the reinjection of some empirical ideas will act to balance purely aesthetic criteria, and it can be said that in the second half of his century this wish was fulfilled. Von Neumann was working during the period of abstract mathematics' maximum expansion; this put forward, as Musil said, "the bold luxury of pure reason," noting that alongside those branches of mathematics supporting physics and engineering there "are immeasurable realms that exist only for the mathematician" (Musil 1990, 41), while society became increasingly closed, oppressive, constrained, and tied to technology. Only when machines entered the realm of logic (or vice versa) did things change, for society and for mathematics: society is becoming lighter, while algorithms, discrete and finite mathematics responded and found their own space. But it is doubtful whether they brought injections of empiricism rather than new worlds, science fiction if not fairy tales.

Not only are the worlds of invisible "bits" in which we are immersed the stuff of science fiction, so are the universes designed by computer models, which are now taken seriously by cosmologists, who claim to recognize their intersections with "our" universe. They are the ones who suggest new plots to the more refined novelists (such as Auster 2017).

At the beginning of the chapter, we cited Thurston's words to advance the suggestion that mathematics is a social product. Some time ago, a heated discussion on the social nature of proofs was triggered by a well-received paper, written by De Millo, Lipton, and Perlis, regarding automatic proofs of correctness in the IT field (De Millo, Lipton, and Perlis 1979); it inspired the so-called humanistic positions, such as those of Reuben Hersh (1927–). Now those theses appear less scandalous, if moderated. We want to clarify that we don't want to support a relativistic position. Fundamentally, the aim of discursive activity is to identify logical relationships—an objective and, in a certain sense, absolute nucleus is recognizable. It's true that every theorem depends on axioms, but the way it depends on them is subordinated to the logic used and the nature of the axioms. If these describe constructions with tools, like those of Euclid, with ruler and compass, then they have a stability rooted in technique that no cultural evolution can modify (at most they can be forgotten). On the contrary, if they have to do with infinity, even though we share Hilbert's idea that no one will be able to drive us out of the paradise opened to us by Cantor, even though we have convincing reasons to accept them, and even though confirmation of their usefulness accumulates in various directions, we cannot rule out revolutionary shifts that will result in a reconsideration. The logic that allows us to prove Euclid's theorems also has a material nature that can almost be mechanized. We don't have to formalize the *Elements'* geometric theory to realize, as we glance over the proofs, that beyond propositional calculus one can equally reduce oneself to using the logical rules of predicative languages, first-order logic (see appendix, 10). It's no coincidence that continuity, which requires another language, is not mentioned (one of the logical gaps noted in the *Elements* is precisely in Proposition I.1, and concerns the asserted intersection of two circles). Able to be mechanized doesn't mean able to be decided, it means enumerable, such that all its theorems are generated by a single mechanism, together with their finite proofs, albeit in an unpredictable order.

Other theories, those considered more significant today, in addition to using axioms that are perhaps more problematic, use a less controllable logic. For example, second-order logic, which presupposes the concept of power $\mathcal{P}$ and which has the advantage of allowing infinite structures to be defined. For this logic, the set of theorems (that is, semantically valid statements) isn't generated by any mechanism. However, mathematicians who work as Thurston describes don't care. The fact that one cannot appeal to a machine that can generate all theorems is irrelevant; they're just interested in one theorem at each moment. Their attitude is that if a statement is interesting and seems to touch on the theory they're studying, a proof will be found, possibly with a special effort; if it requires originality and the invention of new techniques and concepts, even better. This is what's behind the judgment of logic as irrelevant.

But as long as you remain within the framework of set theory, it can be formalized in a language and in a logic of first order. Even

the new theorem that required so much effort will also be reab-
sorbed, perhaps with an appropriate translation (like that of the 3
in {∅, {∅}, {∅, {∅}}}), among the outputs of the set machine. We also
use this logic in automated theorem proving. It's unimaginable
that these tools could become obsolete. It's not because the technique is immortal; think, for example, about how the machines
that the Greeks used to solve the problems of trisecting an angle or
duplicating a cube—condemned by Plato (see section 4.1)—were
forgotten. The reason is that what we call machines could be something more than physical objects, they could be a copy of something
printed on our neuronal circuits, according to the insights of War-
ren S. McCulloch (1898–1969) and Walter Pitts (1923–1969), in
McCulloch and Pitts (1943). If the machines appear to us to oper-
ate in a solid and independent way, it's not because of the heavi-
ness of the hardware, but because they were invented and built by
simulating the physical movements (eyes and hands) connected to
attention and the integrated neuronal logic circuits. And yet imagination is still a mystery.

3

Three Pieces of Evidence

> I would prefer the term *free mathematics* to the customary pure mathematics.
>
> —Georg Cantor

We can present three types of evidence to corroborate the conjecture that the proof is, or derives from, a story. The first is found in linguistics.

3.1 Etymologies

According to the *Oxford English Dictionary*, even today the English "tale" means both story and in some cases "number" or "total." It derives from the Dutch *taal*, which means speech, but is related to the German *Zahl*, number. Similarly, "to tell" can also mean "to determine exactly" and comes from the archaic *tellan*, connected again to the German *zahlen*.

Similarly, in Italian *contare* (to count) is used not only to indicate the act of counting but also "to have a value"; the Italian title of Theodore Melfi's 2016 film *Il diritto di contare* (the right to count) encompasses both. The film's original English title *Hidden Figures* contains another ambivalent term: figures, numbers but also people, shapes, forms. *Contare* is also used to mean *raccontare* (recount), in the sense of telling a story. In many Italian dialects, *lu cuntu* is both the story and *il conto* (the bill). A more extensive investigation into the origin of some words would hold other surprises.

3.2 Dynamic proofs

The history of mathematics provides another type of reflection, with several examples of how mathematicians present their results. As we mentioned in the preface, in mathematics, narrating doesn't mean only telling the story of mathematics; it's also used to explain how an application functions or how a theorem is proved. There is no doubt that creative mathematicians are sensitive to the fact that what they're talking about has to make sense and therefore must be understood, *possibly in the same way that they understood it.* There is a reason that Archimedes (ca. 287–212 BCE) wanted to recount how he achieved his results with the mechanical method and infinitesimals (Archimedes 2013). The method is similar to that which is still used in kindergarten, where a thin sheet of material in the shape of the (geometric) figure is considered, and its weight is compared with that of a shape of a known area, or a bit of cardboard which is then cut into triangles and reassembled. Archimedes replaced these thin shapes, in a thought experiment, with many strips of infinitesimal width—equated to segments—and the shape is conceived as the totality of these segments and balanced against another shape with a known area. The formal proofs that Archimedes added were proved by the method of exhaustion. This method, credited to Eudoxus of Cnidus (408–355 BCE), was a method involving potentially infinite approximations that required finalizing with what we call a convergence proof. The Greeks did it using reductio ad absurdum, assuming a gap between the result of the process and the objective, and thus showing that some approximation was closer to the objective than the gap. Two very different stories.

Archimedes explains that his method confers a certain ease in dealing with mathematical issues that involve mechanical considerations. He wanted this method to be known because he was sure that it would produce other results and would also be useful in view of proving the results. The heuristic arguments are not true proofs—he concedes to Alexandrian orthodoxy—but it is easier to find a proof when one has become familiar with the subject thanks to the information obtained with his method.

Perhaps the Alexandrian ostracism of infinitesimals had mathematical reasons of which we are unaware or, more likely, metaphysical ones with regard to infinity, from the moment that this was rejected by Aristotle as actual infinity. The only infinity is the potential one (ἄπειρον, *apeiron*). Aristotle refers explicitly to geometers, denying that the restriction is an impediment to them: "In point of fact they do not need the infinite and do not use it. They postulate only that the finite straight line may be produced as far as they wish" (*Physics* III.7.207b27–33).

A finite straight line could however be an actual infinity, a continuum. The straight line was a representation of linear space and movement; in fact, the first discussions about infinity are linked to the paradoxes of Zeno of Elea (ca. 490–430 BCE, from the information contained in Plato's *Parmenides*), pupil and lover of Parmenides, inventor of dialectics. Zeno's paradoxes are divided into those rebutting plurality and those trying to demonstrate the impossibility of movement: the arrow which is immobile because in each instant it occupies only one position; Achilles who never overtakes the turtle; the stadium paradox. One of the versions of the latter, lesser-known, paradox—which doesn't refer to the stadium—is presented as follows: given that time is made of instants, if A and B move toward the left and toward the right respectively, one interval in every instant, after one instant they are two intervals apart; but then, one half of an instant would be necessary for them to be one interval apart.

Aristotle discusses the continuum (τὸ συνεχές) with detailed definitions in part III of book V of *Physics*, although he does so using terminology not immediately interpretable: two terms in a series are called consecutive if there is no intermediate term of the same type; they are contiguous (ἐχόμενον) if consecutive and in contact, that is if the extremities (τὰ ἄχρα) are together (ἅμα), which is to say in the same place.

For Aristotle, the continuum is a kind of contiguous, to characterize which he says that between two consecutive terms the boundary is one. Two consecutive parts are held together (συνέχω), which is not possible if there are two boundaries rather than a single shared one. To understand what he means, it should be noted that by dealing with what is intermediate, Aristotle had argued that what changes does so according to nature and with continuity (συνεχῶς) in the sense that, while the opposite is the ultimate terminus of the change, before reaching the opposite that which is changing reaches that which is intermediate; change and intermediate are connected to the continuum. Continuous is said of a thing if the limit (πέρας, *peras*) that exists between it and its modification is unique.

So, according to Aristotle, who is essentially thinking about the straight line, the continuum is something that changes with continuity. It cannot be made of indivisibles (ἐξ ἀδιαιρέων). The continuum is that which is always divisible into parts which are always divisible, never into indivisible parts (*Physics*, book VI). This implies that the continuum is not, as it is now for us, a set of points, but rather a set of intervals (today the intuitionist concept of the continuum is roughly similar). For Aristotle, points are the extremities of an interval; two points cannot be either contiguous or continuous, as they have no "ends"; the end is other than what it is the end of, and points don't have parts. The points on the line, Aristotle replies to Zeno, are only potential, not actual.

But isn't it surprising that infinitesimals have continued to reappear in history, from Archimedes to Cavalieri, to Newton, to Cauchy, and even today, despite the difficulty of dealing with them rigorously, always with the justification of their simplicity and their ability to favor the discovery not only of results but also of their proofs?

In his time, Bonaventura Cavalieri was considered the true heir of Archimedes, by Galileo among others, both for his interests and for his acuteness and methods. He took up Archimedes' idea along with his indivisibles, conceiving them dynamically: inviting the consideration of two parallel tangents to a plane figure, thought of as the intersection of two parallel planes with the plane of the figure; then imagining that one of these parallel planes moves toward the other, staying parallel to it, until they coincide and the lines where the parallel plane in motion intersected the given figure were the indivisibles, the sum of which was identified with the figure. That totality was called (ambiguously, because it is not clear if the expression refers to the whole or all of the individual lines in a distributive way) "all of the lines of the figure" (*omnes lineae figurae*) and was compared with all of the lines of other figures; similarly for solids and plane sections. In Archimedes' wake, in reasoning about "all the lines," Cavalieri calculated areas and volumes; importantly, the areas subtended by polynomial curves. Today, in static set theory language, we would say that the figure is the totality of segments intercepted on the figure by a bundle of parallel lines. Then, a bundle of parallel lines was conceived as a result of the parallel movement of a transverse plane and its intersections.

Dynamic and anthropomorphic terminology is typical of presentations that want to make inspiring ideas understandable.

With the work of Leonhard Euler, the infinitesimals become numbers, albeit with the difficulty of describing them coherently, at least until the solution found by mathematical logic in the twentieth century. But Cauchy, the first to define the infinitesimals, introduced them as sequences (or functions), rather than as numbers. His infinitesimals are in motion.

> When the successive numerical values [positive, as numbers, we would say absolute values] of such a variable decrease indefinitely, in such a way as to fall below any given number, this variable becomes what we call *infinitesimal*, or an *infinitely small quantity*. (Cauchy 1897, 19; 1899a, 23; 1899b)

With this definition, together with that of limit, the first modern one, Cauchy began bringing rigor to nineteenth-century analysis. According to the definition, infinitesimals are variables that tend to zero. In his proofs, Cauchy continued to use the language of infinitesimals and (more rarely) of infinities. In the *Avertissements* of his 1823 and 1829 texts, he added the clarification: "my principal aim has been to reconcile the rigor which I had made a law for myself in my *Cours d'analyse*, with the simplicity that results from the direct considerations of infinitesimals" (Cauchy 1899, 9; English translation from Cauchy 2013, 124). However, "simplicity" is a vague, relative, and entirely personal idea; it seems more probable that contextualizing the narrative in the language of infinitesimals would allow Cauchy to express the meaning of theorems in a way that seemed simpler to him, which is to say in the way he himself had thought of it. The use of infinitesimal language is held accountable for what became known as "Cauchy errors," although it is possible that the interpreters do not grasp Cauchy's particular way of reasoning. Among the often-cited errors is the theorem that a convergent sequence of continuous functions in an interval tends to a continuous function, and the one according to which a continuous function in a finite interval admits the integral. Lolli (2017, ch. 9.8) summarizes the arguments of Detlef Laugwitz (1932–2000) who invites a reinterpretation of Cauchy in terms of his concepts. It would then appear that these did not coincide perfectly with ours; for example, Cauchy did not define continuity in a point but only around a point.

The examples of dynamic terminology could be multiplied. When John Napier (1550–1617) invented logarithms in 1614, he didn't invent them as they are now taught in schools (those are the refinements of Henry Briggs (1561–1630)). Napier imagined two straight lines, the first divided into intervals of equal length, the second a segment of length r divided by a sequence of points whose distances from the far right formed a decreasing geometric sequence:

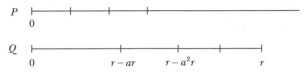

The distances were ar, a^2r, a^3r, ... In fact, Napier used $r = 10^7$ and $a = 1 - 10^{-7}$ for his tables.

Napier saw two points moving on the two segments at different speeds, one at constant speed and one at a speed decreasing in geometric proportion, as a function of the distance from the end of the segment, and connected the positions thus reached within the same times. To simplify the calculations of his tables, which resulted in numbers that were too small, he inserted a corrective factor. For this reason it was no longer true that the logarithm of the product was the sum of the logarithms: so his logarithms were only approximate. This disproves the cliché that the aim of mathematics is precision. The aim is knowledge; precision is an added bonus that comes out of later improvements.

Alan Turing (1912–1954) resorted to the use of certain machines of his to define algorithms. Other definitions were of the arithmetical type (class of functions closed with respect to operators such as recursion) or logical (deductions in the calculus of equations or in λ-calculus). With Turing, explaining an algorithm becomes the story of a machine's actions: moving its parts, observing what it has done, wondering how to proceed, and deciding on the basis of the instructions it reads. The language is not teleological, only analogical. That the machine reads a symbol means that (as we already mentioned in section 2.4) at a given instant, a pointer is positioned in front of a box where that symbol is inscribed and the machine reacts differently to different symbols (depending on its current state and the instructions it consults); that the machine questions itself means that the list of instructions is scanned to recognize a coincidence of symbols. The story introduces movement, the making of a design to obtain the result, "thinking with the hands" as Primo Levi would have said.

The fascination of movement's language is still evident today in the attraction exercised by categorical language; one quality that has made it so popular is that its main concept, morphism, expresses a transformation, a transport of structure.

3.3   The nature of mathematical entities

A third type of evidence can be recognized in reflecting on the nature of mathematical entities, as they are conceived by mathematicians, not in a philosophical context. A particularly pertinent notion is that of the fictitious or "ideal elements" frequently used in modern mathematics. Gottfried W. Leibniz (1646–1716) managed to invent an efficient calculation that avoided the question of what dx and dy were, which actually are infinitesimals. Leibniz tried several times to explain the methods of infinitesimal calculus, first of all to himself. There are several vacillations, but it can be said that he firmly kept these principles: that infinites and infinitesimals do not have a metaphysical reality, that they are a useful fiction to shorten the reasoning process and facilitate discovery; however, if you want you can do without them as the ancients did (with the method of exhaustion).

After having observed that infinity only means the excessively large, that, for example, the sun's rays are considered parallel in that they come from very far away (it is said "they meet at infinity"), he concludes:

> even if someone refuses to admit infinite and infinitesimal lines in a rigorous metaphysical sense and as real things, he can still use them with confidence as ideal concepts (*notions idéales*) which shorten his reasoning, similar to what we call imaginary roots in the ordinary algebra, for example, $\sqrt{-2}$. (Leibniz 1969, 543)

Speaking of the imaginary, Leonhard Euler observed that the square roots of negative numbers can be neither negative, nor positive, nor null, and are thus impossible, and "therefore they are usually called *imaginary quantities*, because they exist merely in the imagination." However, these numbers present themselves to our minds, we have a sufficient idea of them and "nothing prevents us from making use of these imaginary numbers, and employing them in calculation" (Euler 1984, 43).

David Hilbert will repeat Leibniz's arguments on infinity as a large finite, and not one existing in nature, almost to the letter. Like Leibniz, he will speak of "ideal elements," a term that had already entered the mathematical lexicon of his time not only for the points at infinity but also for Dedekind's ideal numbers. On this subject he recalled: "These ideal 'infinite' elements [in geometry] have the advantage of making the system of connection laws as simple and perspicuous as possible" (Hilbert 1983, 5).

However, Hilbert recognized in the hitherto episodic appearance of ideal entities the signal of a fundamental characteristic of mathematical thought: "Let us remember that we are mathematicians and that as mathematicians we have often been in precarious situations from which we have been rescued by the ingenious method of ideal elements" (Hilbert 1983, 9). For more information see Lolli (2016a). In section 1.4 we noted that Hilbert's program proved unworkable, in light of the incompleteness theorems for sufficiently strong theories.

Dissatisfied with the solutions to tame infinity offered by the logic of the time, he had the idea of not only supporting but also proving with mathematical certainty that infinity can be considered an ideal element, with an economic and unifying function similar to that already explored with points at infinity. For this outcome, he would have had to prove the noncontradictory nature of infinity theory with finitary tools. The infinite would then be convenient to simplify proofs, shorten them, but in reference to concrete objects it would not demonstrate anything new, that is, it would be a conservative extension of finitary thought. At the same time, it's no longer possible to do without infinity because mathematical research primarily takes place in infinite worlds. In the commonly accepted ZF set theory, one of the axioms states that an infinite set exists (in section 1.2 this axiom was called the "big bang" of universe V).

Infinity would then be relegated solely to the role of idea, "if one means by an idea, in Kant's terminology, a concept of reason which transcends all experience and which completes the concrete as a totality" (Hilbert 1983, 14) (remembering, given the reference to Kant, that for the latter, reason is not the intellect but rather the faculty with which we try to look beyond experience)—an idea in which, thanks to the theory conceived by Hilbert, one could have unhesitating confidence.

Therefore, the concept of fiction, which replaces the impossible, giving it logical license, has some place in mathematics. But why are some entities called fictitious, imaginary, ideal? With this term Leibniz wanted to say that infinitesimals don't have metaphysical reality; but, now that in pure mathematics we don't care about the metaphysical solidity of entities, we recognize that the new entities were only fictitious relative to the frame constituted by the universe of those that were accepted in standard practice. In reality they are all cut from the same cloth, presented axiomatically, even if introduced with the most varied intuitions, fantasies, motivations, and justifications. Then is it not natural to think that all mathematical objects are fictitious?

The transformation of fictitious into normal isn't the result of a sudden, convenient or subjective decision, but rather a gestation involving experience, reasoning, and beliefs. This was noted by the Italian novelist Alessandro Manzoni (1785–1873) in reference to a completely different topic—changes in jurisprudence, particularly in regard to torture:

> In great matters as in small, there comes a moment when that which, being accidental and *fictitious* [*fittizio*], wants to be perpetuated as *natural* and necessary, must at last succumb to experience, reasoning, society, or fashion—or to something even more trivial—according to the nature and importance of the matter in question; but that moment

must be prepared for. (Manzoni 2014, 1162–1163, italics mine; note that in the original Italian text the word *fittizio* could be translated as either "fictitious" or "artificial")

The preparation, in the case of mathematical entities, consists in the exploration of relationships with other, already familiar entities, in ascertaining the possibility of a harmonious coexistence, in seeking consistency as Georg Cantor understood it, as compatibility with previously accepted and established concepts. His infinite numbers are added to the usual numbers if it can be done, and it can, without destroying the accepted certainties. According to Cantor, this was impossible for infinitesimals. Their existence would have contradicted what is known as the Archimedean property, the fact that given two numbers $0 < a < b$ there exists a multiple $n \cdot a$ of a such that $b \leq n \cdot a$: it is the property of the numeric system that allows us to recognize that integers are cofinal with real numbers, that is, for every real number there is a larger integer. Cantor was also convinced that if an assertion was incompatible with a theory T, it would have been refutable by T (essentially assuming the completeness of the theories, at least of that of real numbers; see Lolli 2017), and he made every effort to do this for the Archimedean property, against those who proposed axiomatic theories of non-Archimedean fields. It was a fool's errand in which he was followed by Giuseppe Peano who vainly tried to endorse and complete his proof of the impossibility of infinitesimals. Cantor, although for him freedom was the essence of mathematics, still had a cumulative idea of mathematical heritage and didn't accept the reasons of the axiomatic method with its multiplicity of interpretations.

4

Tradition

I cannot yet tell . . . but I think, if I may put it so, that I have a certain helpful hypothesis for the problem.
—Plato

4.1 Plato and Aristotle

We know that, for Plato, mathematics was made up of timeless truths, or truths that existed outside of time, "the knowledge of the eternally existent" (Plato, *Republic* 527b). We've seen in section 2.4 how, according to the anamnesis theory, geometric knowledge was recalled by the memory when activated by the observation of figures. And consequently, Plato consistently protested against the way those who practiced geometry described it:

> "This at least," said I, "will not be disputed by those who have even a slight acquaintance with geometry, that this science is in direct contradiction with the language employed in it by its adepts. . . . Their language is most ludicrous, though they cannot help it, for they speak as if they were doing something and as if all their words were directed towards action. For all their talk is of squaring and applying and adding and the like." (Plato, *Republic* 527ab)

Who are these practitioners of geometry? They must have been geometers, but Plato had idiosyncrasies, and he changed his opinion over the course of his life. In his first dialogues he appreciated the geometers' procedures and analytical method; the way they traced a problem or a statement back to simpler problems or statements that they knew how to resolve or decide. From this quotation, together with the next one from Plutarch, it seems that Plato meant that those who practice geometry are mathematicians who want to use mechanical means to force propositions for which they lack proof. Mechanical instruments appeared in the research of those who were dealing with problems that couldn't be resolved with a straightedge and compass—like trisection of the angle and duplication of the cube—which required the drawing and intersection of higher-order curves. A curious and ironic circumstance is that, traditionally, Plato himself was associated with the invention of such an instrument (a doubtful attribution, supported only by Eutocius, a sixth-century Byzantine mathematician; see the discussion in Heath 1981, 1:255–258).

Hippocrates of Chios (470–410 BCE) had shown that doubling the cube could be achieved by finding two lengths that are mean proportionals in continuous progression between two lengths, one of which is double the other: $a:x=x:y=y:b$, and Menaechmus (380–320 BCE) observed that since it had to be $x^2=ay$, $y^2=bx$, $xy=ab$ the problem was reduced to finding the intersection of two conics:

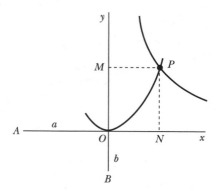

Two mean proportionals between AO and OB.

The solution could be constructed with this instrument (how it works is explained in Heath (1981, 1:256–257):

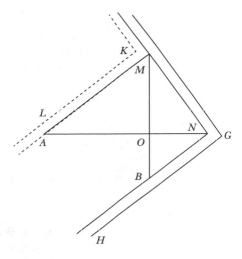

The instrument attributed to Plato.

In his last writings, from which the quote from the *Republic* was taken, Plato criticizes the geometers' way of sheathing the results in axiomatic armor. He reproached them, pointing out the limitations, compared to dialectic, of stopping at the axioms in the backward ascent in the search for knowledge.

His invective was quite successful then, and had consequences that are still being felt. The following passage from Plutarch (46–120 CE) is well known:

> For the art of mechanics, now so celebrated and admired, was first originated by Eudoxus and Archytas [430–365] . . . and gave to problems incapable of proof by word and diagram, a support derived from mechanical illustrations that were patent to the senses. For instance, in solving the problem of finding two mean proportional lines, a necessary requisite for many geometrical figures, both mathematicians had recourse to mechanical arrangements, adapting to their purposes certain intermediate portions of curved lines and

sections. But Plato was incensed at this, and inveighed against them as corrupters and destroyers of the pure excellence of geometry, which thus turned her back upon the incorporeal things of abstract thought and descended to the things of sense, making use, moreover, of objects which required much mean and manual labour. For this reason mechanics was made entirely distinct from geometry, and being for a long time ignored by philosophers, came to be regarded as one of the military arts. (Plutarch, *Marcellus*, in Plutarch 1955, 5:471–473)

According to Geoffrey E. R. Lloyd (1933–), Aristotle also considered mathematics as focused on abstract elements, which certainly do not change over time and therefore wouldn't be comparable to a narrative, which in *Poetics* 7.1450b26–27 is considered an account of a chronological sequence of actions with a beginning, a middle, and an end. All the same, Aristotle recognizes, as Lloyd (2012) explains, a dynamic that brings actualization into proofs through action.

The following quotation refers to two theorems which we discussed in section 2.1, the first about the sum of the internal angles of a triangle and the second according to which, if an angle with a vertex on a circle is subtended by a semicircle, then it is a right angle. The interpretation of the passage is challenging due to the polysemy of different terms: *energheia* means both "activity" and the "actualization" of certain properties that exist potentially; *diagrammata* are the diagrams, the geometric propositions, and the proofs as well.

Geometrical constructions [διάγραμματα, diagrams, proofs], too, are discovered by an actualization [ἐνέργεια, *energheia*], because it is by dividing [διαίρεσις] that we [mathematicians] discover them. If the division were already done, they would be obvious; but as it is the division is only there potentially. Why is the sum of the interior angles of a triangle equal to two right angles? Because the angles [of a juxtaposed triangle] about one point [in a straight line] are equal to two right angles. If the line parallel to the side had been already drawn upward, the answer would have been obvious at sight. Why is the angle in a semicircle always a right angle? If three lines are equal, the two forming the base, and the one set upright [perpendicular] from the middle of the base, the answer is obvious to one who knows

the former proposition. Thus it is evident that the potential constructions are discovered by being actualized. The reason for this is that
the actualization is an act of thinking. (*Metaphysics* 1051a21–31)

The term *energheia* should be distinguished from *kinesis*
[κίνησις], movement, modification; Aristotle is certainly not thinking of physical movements, actuality is brought into being through
the exercise of thought. In the text *diagrammata* means proofs,
because these are discovered through constructions, and it would be
tautological to say that the constructions become evident due to the
constructions.

Energheia is complete in every moment, we see it immediately.
But in mathematics one can speak of discovering a truth through
constructions; these apply to abstract entities, which as such are not
altered, yet the constructions actualize what exists potentially.

In Aristotle's passage about the second theorem mentioned,
there is some difficulty in the translation of the term *orthe* (ὀϱθή),
here translated as upright (perpendicular), which would lead to a
construction valid only for an isosceles right triangle, if left as is
without additional constructions. However, it would need to be
completed with Euclid's Proposition III.21 which demonstrates that
in a circle the angles in the same segment are equal to one another.
Or *orthe* can be translated simply as "line"; then there would be a
slightly more complicated proof where the segments forming an
isosceles triangle must be produced (as seen in section 2.1). The
proof would then be wholly general, as in Euclid's figure:

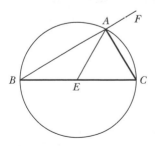

As for the other theorem cited, on the sum of the internal angles of a triangle, Proposition I.32 in the *Elements*, the fact that we're talking about a segment drawn upward suggests that Aristotle was thinking of Euclid's proof based on the following drawing and not that of Eudemus where a straight line parallel to the base is drawn in *A*.

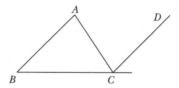

In this figure, for the properties of parallels cut by a transversal the angle at *A* is equal to the angle $A\hat{C}D$, and that at *B* to the angle formed by *CD* and the extension of the base, so the sum of the three internal angles is equal to the straight angle at *C*.

4.2   Modern heirs

The two different legacies of Plato and Aristotle run through the entire history of western thought and are still recognized today. Despite their infinite variations, philosophies of mathematics are classified in two categories according to whether they are influenced by Plato or by Aristotle.

For example, Godfrey H. Hardy (1877–1947), Marc Kac (1914–1984), and Stanislav Ulam (1909–1984) all speak of proofs as "seeing," but in very different ways. Hardy declares:

> I have myself always thought of a mathematician as in the first instance an *observer*, a man who gazes at a distant range of mountains and notes down his observations. His object is simply to distinguish clearly and notify to others as many different peaks as he can. (Hardy 1929, reprinted in Ewald 1996, 1257)

Hardy doesn't feel himself to be an observer-explorer but rather a contemplative observer. A lot of water has passed under the bridge since Stevin's time; pure mathematics no longer seeks the secrets of nature but rather contemplates imaginary worlds. Yet precisely for this reason, seeing through the eyes of the imagination would suggest a narrative, at least the metaphor of climbing up to the observation point, necessarily up high. Hardy immediately realizes that his phrase is exaggerated, if taken literally it would imply that there is nothing to call "proof," but believes that it can be accepted as almost faithful to the practice. He probably intends to allude to the sense of objectivity that emanates from the external projection of what is intuited, in a rigorous language, for example that of set theory. But it is precarious to build a definition of proof on these subjective impressions. Indeed the justification of the necessity and function of proofs is the sore point of mathematical Platonism.

By contrast, Kac and Ulam seem like two Aristotelians. It's no coincidence that they emphasize the use of mechanical instruments (condemned by Plato) to perform operations or graphic representations of curves (Kac and Ulam 1992, 62). According to them, "The art of mathematical proof often consists in finding a framework in which what one is trying to prove becomes nearly obvious" (this statement is also attributed to Herbert A. Simon (1916–2001), regarding solutions to problems). Mathematical creativity manifests itself to a large extent in knowing how to find such frameworks; it sometimes happens that what the mathematician is interested in fits into an already existing frame, invented for a completely different purpose, but sometimes, and this is the highest form of creativity, the mathematician invents such frames (ibid.). The division process Aristotle describes, commenting on the theorem mentioned above, is no different from the construction of a framework, albeit in the same language. It highlights all the elements that participate in the result, like people invited to a party who come to pay homage to the guest of honor. In other cases a change is required, locating or constructing a new framework, which can be a different structure superimposed on that of the problem's terms. Kac and

Ulam's example is Wilson's theorem, where an algebraic structure is superimposed on the numerical one.

The theorem states: If p is a prime number, then $(p-1)!+1$ is divisible by p. For the proof, we consider the residue classes modulo p, that is the set $\{1, 2, \ldots, p-1\}$ with the multiplication $\circ$ defined by $i \circ j =$ the remainder of the division of ij by p, which form a commutative group. One proves that $(p-1)!^2 = 1$, associating every element to its inverse in the product $(p-1)!\,(p-1)!$ If k is an element of the group such that $k \circ k = 1$, then $k^2 - 1$, which is the product of $(k-1) \circ (k+1)$, is divisible by p, and k is either 1 or $p-1$. If in the product $1 \circ 2 \circ \cdots \circ (p-1)$ we pair off and simplify each element with its inverse, $p-1$ remains, the only element other than 1 that is equal to its inverse. Therefore $1 \circ 2 \circ \cdots \circ (p-1) = p-1$, which is equivalent, in a somewhat veiled form, to the statement of the theorem.   □

But once we shift over into group language, other ideas come up. In finite group theory one of the first results is that in every finite group with n elements, every element multiplied by itself n times is equal to the unity. In the group of residue classes, we have that if $1 \le k \le p-1$ then $k \circ k \circ \ldots \circ k = 1$, that is, $k^{p-1} - 1$ is divisible by p, which is Fermat's little theorem.   □

5

Mathematical Language

> Arithmetical symbols are written diagrams and the geometrical figures
> are graphic formulae.
> —David Hilbert

5.1 Symbols and meanings

Remember that in the modern conception of the proof, tied to the
axiomatic method, it's advisable not to think about the meaning of
the terms that appear within it; if that were necessary, it would be a
defect of the deduction or the assumptions. It has also been pointed
out that, even in fairy tales, the meaning of the terms is ignored
with the exception of some almost axiomatically defined characters. However, the reader who is well aware that sometimes entire
pages of a mathematics text are covered in symbols has reason to
think that mathematical discourse is entirely symbolic.

Etymologically, "symbol" derives from the Greek σύν (*syn*) and
βάλλω (*ballo*), which means "throw together." Originally, the two
halves of a card or other object divided in two had to be put together
to verify the identity and rights of two parties to a pact. Later the
symbol puts together, connects, a material mark with an object, in
the broad sense, often identified with a mental image or a super-
sensible reality. The meaning can be conventional or analogical.

Philosophical and anthropological literature dealing with the symbol is extensive but not very relevant for mathematics. For example, Ernst Cassirer (1874–1945) came to the conclusion that a symbol manifests a multiplicity of meanings, in comparison with the sign, and therefore cannot be used in abstract logic (Cassirer 1925). On the contrary, this characteristic represents its added value in logic. However, in mathematics all symbols that have been gradually introduced, other than the letters of the alphabet, are more modestly called "notation," as if they were letters of the mathematical alphabet (see Cajori 1993). The word "sign" is not used much, as it seems too generic. "Symbol" is used (was used), for example, in "symbolic logic" to indicate formal logic, alluding to the entirely artificial nature of alphabets; Peano called it "ideography," believing that symbols represented ideas. When we say that symbols are signs, we risk implicitly assuming that they have no meaning, in a formalist sense.

Hilbert had intimately experienced this ambiguity when he first launched his program. In describing a first draft of the finitist methods, Hilbert accepted the Kantian hypothesis of a prerequisite for the functioning of logic, an a priori capacity, reducing it however to the mastery of manipulations of discrete objects which are experienced "before all thought"; such objects must be "capable of being completely surveyed in all their parts," and "their presentation, their difference, their succession (like the objects themselves) must exist for us immediately, intuitively, as something that cannot be reduced to something else" (Hilbert 1922; English translation in Ewald 1996, 1121). These finite objects were the signs, therefore "the objects of number theory are for me . . . the signs [*Zeichen*] themselves whose shape [*Gestalt*] can be generally and certainly recognized by us—independently of space and time, of the special conditions of the production of the sign, and of insignificant differences in the finished product" (ibid.). The signs, such as 1 or |, or |||, or || . . . |, "which are numbers and which completely make up the numbers" are the object of our consideration, "but otherwise they have no *meaning* [*Bedeutung*] of any sort" (ibid., 1122). With the intention of excluding from logic "abstract operation with general concept-scopes and contents," which in Frege's and Russell's thought had not led to any solution but rather "proved to be inadequate and uncertain," Hilbert ended up making a statement for which he was much reproached: "The solid philosophical attitude that I think is required for the grounding of pure mathematics . . . is this: *In the beginning was the sign*" (ibid.). However, in his later writings, also in order to avoid controversy over the statement "sign without meaning," the term "number sign" was replaced by "numeral."

But for a long time the problem of the meaning of signs had not arisen, at least not in a way that interfered with the mathematicians' work. Certainly it isn't the form of the signs that is significant, which depends on various factors, from the abbreviation of words, initially perhaps in other languages (like Peano's ε from the first letter in ἐστί, "is" in Greek), to typographical necessities, which are present in Descartes and also important for Peano.

In Euclid AB denoted

and $B\hat{A}C$ the angle

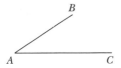

and so on.

The + sign indicated numerical addition, at least from the formation of algebraic language in the modern age onward; it represented the shift from syncopated algebra to symbolic algebra; and similarly other notation. The + and − signs appear in Germany during the last 20 years of the fifteenth century. The · sign for multiplication was introduced by Leibniz in 1698 because × (used since 1631) was easily confused with x. The $=$ sign made its debut in 1557, but there were initially difficulties because it was used in other ways (for example by Viète to express difference). Descartes used ∞, according to some as a deformation of *æ*, the first letters of *aequalis*, and according to others it was the astronomical symbol for Taurus turned on its side, because that was available at printers' shops, where many astronomy texts were published.

As soon as theories with more than one interpretation began to be taken seriously, the problem of the meaning of these symbols arose. In English symbolic algebra, George Boole (1815–1864) still defined a "sign" as any "arbitrary mark, having a fixed interpretation." George Peacock (1791–1858) defined algebra as "a science of symbols and their combinations, constructed upon its own rules, which may be applied to arithmetic and to all other sciences by interpretation" (Peacock 1833, 194–195). For more information on the importance of the discussion regarding symbols for the definition of the axiomatic method, see Lolli (2016a, ch. 1).

There are fluctuations between the various authors: the interpretation is fixed, but also changeable. For Augustus De Morgan (1806–1871), one of Peacock's students, interpretations could lead outside the province of algebra, for example if, in addition to the length of a segment, AB is thought of as a movement from A to B. Interpretation is a change in the meaning of words, which indeed have a meaning, but which one may agree to transpose. It was in fact De Morgan who, as a joke, invented what became the very popular phrase: "sound void of sense."

However, Peacock distinguished between arithmetic algebra, where $+$ and $\cdot$ have their usual meaning of addition and multiplication, and symbolic algebra in which the rules found in the first interpretation are maintained but their applications are extended to all possible values of the symbols. The English algebraists weren't formalists, they didn't claim that the symbols had no meaning and should be manipulated blindly. De Morgan meant that one doesn't think about the meaning, and other proponents of the axiomatic method thought as he did, including Pasch.

For Boole, the interpretation of signs didn't necessarily have to be unique. He observed that recognizing the possibility of multiple interpretations had been delayed by the fact that, historically, the elements to be determined had almost always been conceived as measurable—with ideas about size or numerical relationship predominating. Boole's particular interpretation of algebraic symbols had nothing to do with magnitudes but rather with classes, conceived or recognized by the mind; to the laws of ordinary algebra, he then added the request that the law $x^2 = x$ be satisfied (Boole 1847, 4).

From the experience of the English algebraists, explicating some of Peacock's remarks, Boole deduced the moral that "the meaning of the operations performed, as well as the results obtained under such circumstance, must be derived from the assumed rules, and not from their definitions or assumed meanings." Hilbert is clearly inspired by this moral.

5.2 The rules of the formula game

In 1900, in the introduction to his presentation of mathematical problems, Hilbert reminds us that wherever new concepts emerge, from gnoseology, from geometry, from the natural sciences, the task of mathematics is to investigate the principles that underlie these concepts.

> To new concepts correspond, necessarily, new signs. These we choose in such a way that they remind us of the phenomena which were the occasion for the formation of the new concepts. So the geometrical figures are signs or mnemonic symbols of space intuition and are used as such by all mathematicians. Who does not always use along with the double inequality $a > b > c$ the picture of three points following one another on a straight line as the geometrical picture of the idea "between"? (Hilbert 1900; English translation in Ewald 1996, 2:1100)

Arithmetical symbols, Hilbert states with a powerful metaphor, are written figures and geometric figures are drawn formulas. And no mathematician could do without these drawn formulas, any more than in calculation he or she could dispense with the insertion and removal of brackets, or the use of other analytical signs. But:

> The use of geometrical signs as a means of strict proof presupposes the exact knowledge and complete mastery of the axioms which underlie those figures; and in order that these geometrical figures may be incorporated in the general treasure of mathematical signs, there is necessary a rigorous axiomatic investigation of their conceptual content. (ibid., 2:1100–1101)

In Hilbert's analysis, the fact that the signs don't have meaning doesn't mean that they don't make sense. The sense of the signs lies in the logical analysis of the initial intuition of a new concept and is expressed in the axioms that mathematically define that concept. Every use of them is the development of a deduction from the concept.

Later on, once the program started, Hilbert considered the complete formalization of the higher theories, reducing them to sets of symbols and strings of symbols. In order to formalize the proofs, the further step of introducing logical symbols, connectives, and quantifiers was needed, in the same way as algebraic symbols. Hilbert never returned to the subject of the relationship between intuition and the axiomatic regulation of symbols, except for what concerns logical symbols. Happily for Hilbert, there was already something called logical calculus: "we find the logical calculus already worked out in advance. To be sure, the logical calculus was originally developed from an altogether different point of view. The symbols of the logical calculus originally were introduced only in order to communicate." Now logical symbols, as was already the case with mathematical symbols, can also be subjected to the same transformation, so that the formulas of logical calculus in and of themselves don't mean anything, have no content (which would be doubtful, if they are dealing with infinity) but are ideal statements. Axiomatic analysis of logical operations translates into rules and axioms for logical symbols.

> The rules whereby the formulas are derived from one another correspond to material deduction. Material deduction is thus replaced by a formal procedure governed by rules. The rigorous transition from a naïve to a formal treatment is effected, therefore, both for the axioms (which, though originally viewed naïvely as basic truths, have been long treated in modern axiomatics as mere relations between concepts) and for the logical calculus (which originally was supposed to be merely a different language). (Hilbert 1983, 197)

For example the symbol $\rightarrow$, called the conditional, in itself does not have the meaning of implication, it has no meaning; but its use is bound (in natural deduction calculus) by two rules: one of elimination, so from $A \rightarrow B$ and A one can pass to B, and one of introduction, so if B is deduced from A, $A \rightarrow B$ can be added, eliminating A from the assumptions. The rules specify *how* we treat the implication through the conditional, so they provide the meaning of the implication.

Of his proof theory (*Beweistheorie*, another word for metamathematics used by Hilbert) he states that the fundamental idea is "to describe the activity of our understanding, to make a protocol of the rules according to which our thinking actually proceeds. Thinking, it so happens, parallels speaking and writing: we form statements and place them one behind another" (Hilbert 1996, 475). This part of Hilbert's program is not transitory.

His opponent, Brouwer, in his 1907 thesis expressed the exact opposite: "Making a mathematical study of linguistic symbols . . . can teach us nothing about mathematics." To Brouwer, who in 1927 had made disparaging judgments about this "game of formulae," Hilbert replied:

> this formula game is carried out according to certain definite rules, in which the *technique of our thinking* is expressed. . . . This formula

game enables us to express the entire thought-content of the science of mathematics in a uniform manner and develop it in such a way that, at the same time, the interconnections between the individual propositions and facts become clear. (ibid.)

Even formal deductions, all written in symbols, express thoughts, not in themselves but thanks to the rules that control the use of logical symbols. So formal exposition isn't an obstacle to the narration of the stories that we conceive.

6

The Origins of Argumentation

There was far more imagination in the head of Archimedes than in that of Homer.

—Voltaire

One of the strongest pieces of evidence is precisely the mysterious beginnings of logic in the Greek world. Literature regarding the origins of proofs is quite extensive.

In recent years many discussions have been catalyzed by the thesis of Árpád Szabó (1913–2001), which supports the idea that deductive mathematics originated in the dialectics of the fifth-century Eleatic school. His principal topic was the abundance of forms of indirect proofs in Euclid's work (Szabó 1978). The discussions were not very instructive, deviating toward controversies between internalists and externalists, because Szabó's thesis seemed to threaten the autonomy of mathematics.

However, a renowned school of classicists, primarily French, has convincingly established that the constitution of those practices which we call "logic" was favored by the needs and opportunities provided by the new Greek political institutions, and participation in city management decisions, in the sixth and fifth centuries BCE (I cite only Vernant 1962 as representative of this school). Democracy as the origin of rationality is especially evident in rhetoric, politics, and law. Scholars have primarily considered the reasoning manifested in natural philosophy as a point of reference, less so that of mathematics. Apostolos Doxiadis, convinced that there is a large degree of crossover between rhetoric and mathematics, advanced the hypothesis of a common ancestor and did a major research study investigating this possibility (Doxiadis 2012). The title of his paper, "A Streetcar Named (among Other Things) Proof," was inspired by a 1947 play of Tennessee Williams (1911–1983), *A Streetcar Named Desire* (made into a film by Elia Kazan in 1951), starting with the observation that "Desire" is a synecdoche (naming a whole after a part), as Desire was the name of only one stop on a streetcar line that was not always called the "Desire Line" but was also sometimes called the "Canal Line" or "Royal Line," the names of other stations on the route. A proof isn't a station, but a line along which there are different stops, a long process. Doxiadis engages in a polite polemic with those who use terms like "birth" or "genesis" to describe the origin of the proof, prompting the idea that, like Athena's gift, they spring fully formed from Aristotle's head. If we have this impression, it is mainly the result of our own ignorance, because there is insufficient documentation on how mathematics was done in Greece before the end of the fourth century BCE.

6.1 From Homer to rhetoric

Doxiadis assumes that readiness to narrate, in the sense of representing actions and states of being, is a basic cognitive ability. He sees in ancient Greece an expansion of the manifestation of this ability that moves from poetic stories to historical narratives to rhetorical argumentation and mathematical proof, in a nonlinear path. Narrative, for example historical narrative, directly affects rhetoric, while its influence on mathematics is mediated by poetic storytelling. In the following pages we briefly summarize the stages of Doxiadis's research and the many themes thoroughly explored and documented in it.

The verbal forms that precede writing are based on parataxis, that is, they have independent clauses concatenated with or without conjunction, or by apposition (a list where consecutive sentences augment the meaning of the preceding ones). The narrative style of ancient stories is still founded on parataxis.

The epics of Homer and Hesiod in the eighth and seventh centuries BCE are the origin of many later narrative forms. The lyric poems of the seventh and sixth centuries are rich in narrative elements.

With tragedy, representation of action rises to a new level through the full exploitation of mimesis (μίμεσις, imitation). Attic tragedy, especially after Aeschylus (525–456 BCE), intersects with rhetoric, and vice versa, in connection with political developments. New forms of speech are aimed at persuasion. There is an increase in the frequency of opinions or general principles, gnomes (γνῶμαι, γνώμη, gnôme), which are already inserted in Homer to indicate analogies with prototypical stories, with paradigms. In the early poets they are taken from popular tradition, like proverbs. Later, new gnomes were invented by poets, and were devised in politics to refer to laws. The paradigms were invoked to justify a rule, followed by an injunction or an invitation. Pericles (495–429 BCE) dispenses with examples and goes directly from rule to injunction.

New forms appear in political oratory, with the goal of persuasion. Quintilian, a Latin orator of the first century CE, who builds on earlier Greek texts, defines narration as the exposition of that which has happened or is supposed to have happened, with the intent to persuade (*Institutio oratoriae* 4.2.31, translated and cited in Doxiadis 2012, 298).

In his *History*, composed between 410 and 400, Thucydides (ca. 460–ca. 399 BCE) exploits the mechanisms of narrative intelligence and theories of the mind: all of the skills necessary to analyze stories in terms of motives and causes. He describes contrasting ideas about political decisions, where opposing sides present narratives of possible futures. Thucydides' art is that of making his own narrative of the future appear necessary, almost like the conclusion of a theorem. He succeeds by limiting his narrative to the events that are directly involved in causal chains (Doxiadis 2012, 299).

Rhetoric comprises three forms: *epideictic*, used for speeches on festive occasions or oratorical exhibitions, which clearly descends from lyric praise poetry; *deliberative* or political; and *forensic*. Even forensic rhetoric relies on convincing representations of actions for its effectiveness, but it transposes many of poetry's artifices into prose. Some expressive forms introduced by poetry to enliven speech proved useful for persuasion and became dominant. Although perhaps initially introduced for aesthetic purposes, those forms became the foundation of the budding method of logical argumentation (ibid., 301).

The new use of narrations aimed not only to represent actions but also to compare narratives, to verify whether they conformed to general rules, the laws of the polis, or to analyze conflicting narratives of events with the aim of convincing a jury of citizens that the version of the speaker is the one that best fits the facts. A new idea arising during the Classical period is that no one-to-one correspondence exists between events and narratives, and that the same events can generate divergent and conflicting narratives. As we see, both Luigi Pirandello (1867–1936), with *Così è se vi pare* (Right you are (if you think so)), and Akira Kurosawa (1910–1998), with *Rashomon*, have ancient forerunners.

The structure of judicial discourse has four parts: the introduction, the narration (πρόθεσις, *prothesis* or διήγησις, *dieghesis*), the proof (πίστις, *pistis*), and the epilogue.

The introduction outlines what is to be proved, in terms of both narration and laws: it was an established strategy of poetic narration, as Aristotle also notes. Both the *Iliad* and the *Odyssey* begin with a summary of what they are going to tell us. The core of the speech is the narration, the pistis only serves to support it. The pistis in turn is usually divided into three parts: division, the pistis proper, and refutation.

The division breaks the narrative down into parts, highlighting the facts that need to be proved or that the adversary can contest. This generates subnarratives and counternarratives. The former expand upon some parts, or question the causes of or reasons for the events.

For example, in the *Encomium of Helen,* from around 430 BCE, Gorgias (485–375 BCE) does not dispute the fact that Helen left and went to Troy, but he asks if this is due to (*a*) the will of the gods or fate, (*b*) a physical constraint, kidnapping, (*c*) successful persuasion, or (*d*) love.

The counternarratives serve to hypothesize alternatives which are dismantled one by one; the arguments used derive from poetic compositions, perhaps following a principle similar to that of exaptation in natural selection: originally the expressions are used for emphasis, later in order to persuade.

The main type of argument takes place through a rhetorical form that can be traced back to Homer: the *priamel* (Latin *preambolum*). The priamel is a series of statements, or examples, presented only to be set aside in favor of the final, winning statement, or superlative example. They're like false objectives of a hunt that ultimately ends with a winner. A later, concise example, from Sappho (ca. 630–ca. 580 BCE):

> Some say that cavalry, some infantry,
> some a fleet, is the most beautiful [thing]
> on the dark earth.
> But I say [it is] one's lover.

We point out this rhetorical form because, as we shall see, it is related to a class of negative logical arguments, such as the reductio ad absurdum.

The pistis had an interesting part focused on witnesses' statements. Since all free citizens, regardless of wealth and status, could be witnesses, there had to be criteria for comparing their testimonies.

The pistis is divided into two parts, one of which—the nonartistic proof—uses material not exclusive to the case in question, such as laws and contracts, brought in if relevant (as we will see in geometric proofs, this would correspond to reference to common notions). The second is the artistic proof, created by the speaker to address the case in question. This in turn is composed of ethos, relating to the moral character of the accused; pathos, aimed at impressing the jury with emotional arguments; and a part called "logic" (*logos pistis*), which originally meant only that these arguments were a part of the overall speech (λόγος), including imported materials.

6.2   The chiasmus

In the framework thus outlined, Doxiadis focuses on two archaic poetry techniques, which Greek speakers used with increasing frequency, the chiasmus and the ring composition. The chiasmus, abbreviated later as X, is a symmetrical structure of sentences that in the minimal form appears as:

$$A\text{-}B\text{-}B^*\text{-}A^*$$

where the asterisk links the phrases through a common or antithetical element (a word, a phrase, a concept) and the common element in the phrase with the asterisk is called the pivot. A classic example is the exhortation of John Fitzgerald Kennedy (1917–1963): "Ask not what your country can do for you—ask what you can do for your country." In this phrase "country" is the outer or A-pivot and "you" the inner or B-pivot.

Rather than two phrases A and B, $2n$ may occur. The first examples are found in Homer, but perhaps in Greece these forms came from earlier literary traditions, such as the Sumerian-Akkadian or

Ugaritic ones. In the *Odyssey* (11.171–203), there is an extended chiasmus when Odysseus speaks with his mother, Anticleia, in Hades (Doxiadis 2012, 314):

  1. What killed you?
    2. A long sickness?
      3. Or Artemis with her arrows?
        4. How is my father?
          5. How is my son?
            6. Are my possessions safe?
              7. Has my wife been faithful?
              7.* Your wife has been faithful.
            6.* Your possessions are safe.
          5.* Your son is thriving.
        4.* Your father is alive but in poor condition.
      3.* Artemis did not kill me with her arrows.
    2.* Nor did a sickness kill me.
  1.* But my longing for you killed me.

The ring composition, abbreviated RC, differs from X in that it contains a single central element: A-B-A* in the simplest case.

The primary reason for the use of X and RC is mnemonic, but it also serves other purposes. These devices also work to structure the narrative by building bridges between the components of the poem and introducing digressive material into the broader narrative. Like many linguistic techniques, these too are best represented in a spatial form: on the other hand, they were also exploited in tragedies, with the movements of the chorus proceeding in one direction during the first part and in the opposite direction during the part marked with asterisks.

The chiastic order, or ring composition, is determined by the arrangement of ideas, to introduce the element of surprise, to suggest one image after another giving the illusion of a synchronic vision and rendering the flow of thought continuous, binding oppositions into a unity.

This ordering has an important function in rhetoric in giving continuity and unity to a series of expressions, even those not connected to the rationale of representing action, perhaps for aesthetic pleasure, but always in view of persuasion (ibid., 317).

The effect of X, like that of other poetic devices, such as rhyme, is similar to the joke, which cannot be reduced to the sum of its parts. Particularly interesting are the quick and powerful forms of X, used to great rhetorical effect in embarrassing the interlocutor. This is exemplified in Gorgias' *Defense of Palamedes*, when Palamedes defends himself against Odysseus' accusation of treason:

    1. If I am wise,
        2. I have not erred.
        2*. If I erred,
    1*. I am not wise.

We find an example of RC in Homer, a paradigm inserted in a mini-story, when Achilles and Priam meet. Remember that Niobe, daughter of Tantalus, had seven sons and seven daughters with the king of Thebes and, proud of her offspring, she made fun of the goddess Latona who had only two children: Apollo and Artemis. To punish her for her pride, Latona ordered her children to kill all of Niobe's. The distraught mother then turned into a block of stone from which a spring flowed.

    1. Let us eat.
        2. For even Niobe ate.
          3. This was her story.
        2*. She ate.
    1*. So let us also eat.

To show a case of demonstration with RC we can look to Lysias' (445–400 BCE) speech "On the Murder of Eratosthenes" (ca. 400 BCE). The speaker killed his wife's lover when he caught them in the act, and Athenian law applied the death penalty to the adulterer. The accused wants to make the judges admit that he himself has carried out what the laws dictated (ibid., 319):

1. Wherefore I, sirs, not only stand acquitted of wrongdoing by the laws

2. but am also directed by them to obtain this satisfaction: it is for you to decide whether the laws are to be valid or of no account.

3. For to my thinking every city makes its laws in order that on any matter which perplexes us we may resort to them and inquire what we have to do.

2*. And so it is they [the laws] who, in cases like the present, exhort the wronged parties to obtain satisfaction.

1*. I call upon you to support their [the laws] opinion.

Lysias's speech is a case of analogy, the next example one of polarity, two conceptual categories in archaic thought. The previously mentioned mini-story, in which Gorgias asks if Helen's abduction was decided by the gods, uses an extended chiasmus:

1. Now if through, it is right for the responsible one to be held responsible

2. for the gods' predetermination cannot be hindered by human predetermination,

3. for it is the nature of things, not for the **strong** to be hindered by the *weak*

3*. but for *the weaker* to be ruled and drawn by the **stronger**.

2*. God is a stronger force than man in might and in wit and in other ways.

1*. If then one must place blame on fate and on a god, one must free Helen [who is not responsible] from disgrace.

In 1 and 1*, there are contrasting elements (responsible, not responsible); in 2 and 2* gods and God are identical pivots of the parallelism; 3 and 3* is, in fact, a gnome in a simple chiasmus, the A-pivot in bold and B-pivot in italics.

Finally, a case where polarity is contradiction, from the *First Tetralogy* of Antiphon (fifth century BCE), an examination of a counternarrative. The counternarrative contains an action, but that action did not occur, therefore the counternarrative did not happen:

1. Nobody who is exposing his life to grave risk [the malefactors]
　2. would forgo the prize [i.e., the cloak] when it was within his grasp.
　2*. And the victim was found wearing his cloak.
1*. Malefactors are not likely to have murdered him.

An examination of the refutation of a counternarrative shows the effectiveness of the chiastic form: the argument in a linear version is shown by a succession of affirmations which, although of the same descriptive nature, are placed at different levels: the counternarrative as it is heard, the observations made, and the conclusion:

$$\text{Counternarrative A} \rightarrow \text{Contains action B} \rightarrow$$
$$\rightarrow \text{B did not occur} \rightarrow \text{A did not happen}$$

The first arrows indicate the assumptions and the subsequent observations, while the last signals the consequence. But the conclusion is not justified solely by the content of the previous phrase but by this one in addition to the preceding ones. A spatial representation of the influence of the previous statements could highlight the difference of the levels involved, as for example in the following diagram:

Story A $\longrightarrow$ A contains B $\longrightarrow$ B does not occur
$\rightarrow A$ did not happen $\leftarrow$

While the nonlinear structure is not expressible in speech, it becomes so with the chiasmus. We will give only one example of the more complex forms, taken from Gorgias' *Defense of Palamedes*.

⎡ 1. I did not commit treason.
│ ⎡ 2. Treasonable action must begin with discussion.
│ │ ⎡ 3. If a discussion occurs, it implies there is a meeting.
│ │ │ 3*a. But a meeting is impossible
│ │ │   4a. nobody came to me
│ │ │   4b. and I couldn't go to others
│ │ │   4c. and I was unable to write and send a message.
│ │ ⎣ (3*b. So there was no discussion.)
│ ⎣ (2*. So no treasonable action.)
⎣ (1*. I did not commit treason.)

In the original the argument is presented in a concise form, 1*-2-3-3*a-4, but according to Doxiadis, those used to the X and RC forms would have no difficulty in creating closure with 3*b–1*, as 1–4 proceed in a cascade.

4a, 4b, and 4c seem to have been added by Doxiadis to enhance the argument, but do not have the character of a gnome, despite being the heart of the defense. They could be inserted in the chiasmus form with a slightly different representation of Gorgias' complete argument:

⎡ 1. I did not commit treason.
│ ⎡ 2. Treasonable action must begin with discussion.
│ │ ⎡ 3. If a discussion occurs, it implies there is a meeting.
│ │ │   4. But
│ │ │   4a. nobody came to me
│ │ │   4b. and I couldn't go to others
│ │ │   4c. and I was unable to write and send a
│ │ │   message.
│ │ ⎣   4*. The meeting was impossible.
│ │   3*. So there was no discussion,
│ ⎣ 2*. so there was no discussion of preparation for the
│   treasonable action.
⎣ 1*. I did not commit treason.

One could do even better by inserting conditions 4*a*, 4*b*, and 4*c* actively in the argument, as in the compound chiasmus.

1. I did not commit treason.
  2. Treasonable action must begin with discussion.
    3. If a discussion occurs, it implies there is a meeting.
      4. But in order for a meeting to take place either someone comes to me or I go to others or I write [a message].
        5. Nobody came.
        5*. Therefore in order for a meeting to take place either I go to others or I write [a message].
          6. I didn't go to others.
          6*. So to arrange a meeting I write [a message].
            7. I didn't write to anyone.
            7*. So, nobody came to me, and I did not go to others, and nobody wrote.
      4*. The meeting was impossible.
    3*. So there was no discussion,
  2*. So there was no discussion of preparation for the treasonable action.
1*. I did not commit treason.

However, the form is found more frequently in mathematical proofs, as we will see.

7

Poetry and Logic in Euclid

Pure mathematics is, in its own way, the poetry of logical ideas.
—Albert Einstein

Finally Doxiadis turns to mathematics to assess the results of his review (refer to Doxiadis and Mazur 2012, 339–371, for a more detailed explication). It is necessary to keep in mind the slow progress of mathematics' material and conceptual tools. For example, on vases from the ninth century BCE we see incised circles and a central point, which likely marks the pivot point of a primitive circle-drawing tool. The adjustable compass will not appear before the end of the sixth century. When a tool invented for a given use is perfected, more sophisticated users invent new uses; this is true of both material and conceptual tools. Doxiadis imagines the first proofs were made by groups of people expert in the use of the new compass around a surface, a table, sand on which to draw. Initially the language was probably deictic: this point, this line . . . As is the case with tools, when the language evolves its uses also become more sophisticated.

7.1   The *Elements*

Doxiadis focuses above all on Euclid, as is natural because his is the only corpus handed down, but the road along the way to Euclid is full of events and characters—starting with Thales of Miletus (ca. 630–550 BCE) and Pythagoras (sixth century), little known due to a lack of documentation.

The first collections of *Elements* begin after the middle of the fifth century, although unfortunately other presentations of this type have remained at most fragmentary, not allowing us to understand how they were organized. Hippocrates of Chios seems to have been the first author of one, but there is nothing left of it, except quotes that mention his results on the squaring of the lunes.

According to some, all of Euclid's first four books are a sometimes ingenious rearrangement of a part of Pythagorean and post-Pythagorean geometry, omitting any consideration involving irrational relationships; Eudoxus of Cnidus' theory of proportions is in fact set out in the fifth book.

The arithmetical part of the following sixth, seventh, and eighth books was largely known by the Pythagoreans, who considered various types of means and had a theory of proportions applicable only to commensurable quantities. This must have been contained in one or more *Elements of Arithmetic* which seem to have existed, at least from the time of Archytas, but have not come down to us. Many results dealing with the plane and solid numbers of Euclid's eighth and ninth books (squares and cubes) were also known. His tenth book, which deals with irrational numbers, was largely covered in the work of Theaetetus (415–369 BCE). Part of the eleventh book, on regular solids, was known, as was much of the thirteenth, once again at least since Theaetetus. The twelfth deals with the method of exhaustion and had not appeared in previous *Elements* in such a systematic way. Eudoxus' contributions were practically the only truly new ones with respect to what Plato knew.

What made Euclid famous over the centuries is the presentation of the first four books, which formed the model and the ideal of mathematical theories. As we said in section 4.1, the axiomatic organization of geometry may have been established just during Plato's lifetime, considering his vacillating judgments on it.

Since Aristotle is a bit earlier than or perhaps contemporary with Euclid, when you compare the two you see what is new in Euclid's *Elements* with respect to the knowledge bank of earlier or contemporary geometers from which Aristotle drew. Aristotle's source of reference is probably an author of a version of *Elements* immediately preceding Euclid's, a certain Theudius.

Aristotle makes use of several geometric proofs to illustrate logical passages, and some differ from those of Euclid. This proves that Euclid did not limit himself to collecting existing results, but revised the proofs so that they fit into his plan.

An important example, and not the only one, is the *pons asinorum*, or Thales theorem on isosceles triangles, Proposition I.5. Aristotle presents a proof of this to illustrate the fact that in any syllogism one of the two premises must be affirmative and universal; however, his proof considers the triangle inscribed in a circle and therefore uses propositions that deal with mixed angles, angles that Euclid only considers in extremely rare cases, and in particular the equality of the angles formed by a chord and the circumference. If before Euclid such propositions occurred in the proof of what, in Euclid, is Proposition I.5, according to some commentators their only possible proof was with some form of superimposition.

Moreover, Aristotle also cites results that are not in the *Elements*: the sum of the exterior angles of any polygon are equal to four right angles, which is also a Pythagorean proposition; the characterization of a circle as the locus of points whose distance from two given points has a constant ratio $\neq 1$; of all closed lines starting from a point, returning to it again, and including a given area, the circumference of a circle is the shortest; the fact that only the pyramid and the cube can fill up space (Heath 1981, 1:340).

Aristotle was also aware of Eudoxus' method of exhaustion, albeit his attitude toward it was critical; some of his considerations on the impossibility of infinite sums perhaps contributed to calling the procedure into question, subsequently requiring Archimedes to go back and explicitly defend it.

The model of the geometric proof was established toward the middle of the fourth century BCE. With occasional variations, it consists of the following steps.

1. The enunciation (πρότασις, *protasis*), where a statement is presented in general terms.

2. The setting out (ἔκθεσις (Ionic), *ekthesis*), where the statement made in the protasis is rendered in a concrete way, by means of examples while maintaining its generality, with reference to a lettered diagram.

3. The specification (διορισμός, *diorismos*), where the statement of the enunciation is repeated in the terms of the setting out, and where the real goal is defined.

4. The construction (κατασκευή, *kataskeuê*), through which points, lines, or shapes that will be useful to the proof are added to the diagram.

5. The proof (ἀπόδειξις, *apodeixis*), which proves the statement as described in 2 and 3, also using the objects constructed in 4.

6. The conclusion (συνπέρασμα, *sumperasma*), or epilogue, which repeats the statement in the general terms of 1.

In Euclid's proofs we find the same types of sentences that are found in the different types of narratives previously examined, atomic sentences that serve as basic building blocks to describe an action or affirm a state of affairs, or expressions of general rules. Doxiadis has examined Proposition I.9 of the *Elements*, where the percentage of different types of sentences is roughly the same as in other, nonmathematical narratives, with action sentences predominating. In Proposition III.7, the percentages are very different, with only one action sentence, two general rules, and all of the rest descriptions of states. The reason for the difference is that the two samples analyzed belong to two different parts of their proofs, the first sample to the part referred to above as the construction, the second to the part that is the proof proper.

The various parts can be divided into two classes: in the first there are protasis, ekthesis, diorismos, apodeixis, and sumperasma, which can be more or less comparable to a narrative, depending on the percentage of state-of-affairs statements they contain; in the second we find the kataskeuê which always has a high degree of similarity with a narrative, since it is made up exclusively of statements describing actions. This explains the difference noted above between I.9 and III.7.

A plea during a court case aims to show that a particular event or set of events occurred as narrated by the speaker, and that certain laws apply to judging these events. The geometric proof aims to show that the concrete situation constructed in ekthesis, diorismos, and kataskeuê involves the truth stated in the protasis.

A comparison can be made between the parts of the geometric proof and the pistis of the rhetorical argument.

1. In both cases division breaks down the problem into possible alternatives and counteralternatives.

2. The proof proper reinforces each of the possible alternatives highlighted in the division.

3. In the legal context, the inartistic proof includes testimonies, contracts, and laws, while in the geometric one it contains references to axioms, definitions, common notions, and already proven theorems.

4. The artistic proof is articulated in the juridical setting thus: ethos presents the moral qualities or character of the person being judged, pathos is based on emotional arguments aimed at the judges, and logos uses only the materials of the narrative and general principles. The artistic proof in the geometric theorem has only the part based on logos, exploiting the materials made available in the ekthesis, diorismos, and kataskeuê.

It's not difficult to recognize the X or RC form in the macrostructure of the proofs:

1. Protasis. The general statement of the theorem: in such and such case (situation), this and this other are valid (result).
    2. Ekthesis. Statement of the situation in reference to a lettered diagram.
        3. Diorismos. Statement of the result, referring to the lettered diagram.
        4. Construction.
        5. Apodeixis, which terminates in 3* and 2*
        3*. End of the apodeixis. Precise statement of the result, which had to be demonstrated or constructed.
    2*. End of the apodeixis. Reaffirmation of the situation referred to in the diagram as in the ekthesis.
1*. Sumperasma. Restatement of the general theorem in terms of the situation and the result.

However, poetic forms also appear in individual microarguments, and are more interesting because they help to express logical inferences.

"Accustomed as we are to observing structures with linear, 'organic' form, we tend to see texts constructed on principles of X/RC as merely formless or messy" (Doxiadis 2012, 346); but the nonrecognition of structure by scholars of a text is precisely the signal that perhaps there is a ring composition within that text, as argued by Mary Douglas (1921–2007), to whom we owe the recognition of the importance of RC outside of philology, as a way of thought, an archaic cultural habit, and a cognitive device (Douglas 2007).

The use of X and RC is conscious and deliberate in Euclid, since it is found even when it is not necessary, for example in the proofs in his *Data*, where simple and straightforward arguments are presented in a convoluted way, one that is seemingly "totally silly" according to Barry Mazur's first reaction (Doxiadis 2012, 371) (we'll return to *Data* later).

The form of RC with a law in the center is very frequent in Euclid. For example in Proposition I.1, with the common notion 1 at the center of the RC:

> 1. Then, CA and CB are each equal to AB.
>> 2. But things equal to the same thing are also equal to one another.
> 1*. Then CA is equal to CB.

A simple binary chiasmus is extracted from I.19:

> 1. If AC is equal to AB,
>> 2. ... the angle ABC is equal to ACB.
>> 2*. The angle ABC is not equal to ACB.
> 1*. Therefore AC is not equal to AB.

The sentence preceding 1 in the proof is "in fact *AC* is not equal to *AB*," which from a division in the rhetorical sense becomes "either *AC* is equal to *AB* or less than it"; therefore the chiasmus takes the first of the two cases "*AC* is equal to *AB*" and refutes it. Later, we will illustrate the proof of I.19 more fully.

We can say the same thing about the binary chiasmus in geometric proofs as we did about the rhetorical chiasmus discussed earlier in relation to Antiphon: it has the ability to represent a complex, nonlinear relationship between different statements of the argument in a linear way.

Note that the structure of modus tollens is as follows:

> 1. If A is true then
>> 2. B is true.
>> 2*. B is not true.
> 1*. So A is not true.

which is related to the negative arguments previously mentioned.

In propositional logic, modus tollens is the rule

$$\frac{A \rightarrow B, \neg B}{\neg A}$$

which is equivalent to the law of contraposition, $(A \rightarrow B) \rightarrow (\neg B \rightarrow \neg A)$ and to the proof via reductio ad absurdum.

The law of contraposition justifies the rule of modus tollens if we admit the cut rule, or modus ponens (see below), which is always present in the calculations: given the assumptions $A \to B$ and $\neg B$, from $(A \to B) \to (\neg B \to \neg A)$, by applying modus ponens twice, we first obtain $\neg B \to \neg A$ and then $\neg A$.

Conversely from the modus tollens rule we have $(A \to B) \wedge \neg B \to \neg A$ with the introduction of $\to$, therefore $(A \to B) \to (\neg B \to \neg A)$ by the law of import-export of the antecedents (i.e., $A \to (B \to C) \leftrightarrow A \wedge B \to C$).

The connection with the reductio ad absurdum proof is less immediate; it requires that the derivability of some laws be established in advance:

$A \vee \neg A$ law of excluded middle;

$\neg\neg A \leftrightarrow A$ law of double negation;

$\neg (A \vee B) \leftrightarrow \neg A \wedge \neg B$ one of De Morgan's two laws;

$A \to A$ law of identity;

$\neg B \to (A \to \neg B)$ law of affirming the consequent.

From these, the law of reductio ad absurdum $(A \to B \wedge \neg B) \to \neg A$ is proved with modus tollens, after observing that $\neg (B \wedge \neg B) \leftrightarrow B \vee \neg B$; therefore $\neg (B \wedge \neg B)$ is a law, the law of noncontradiction:

$$\frac{A \to B \wedge \neg B \quad \neg (B \wedge \neg B)}{\neg A}.$$

The weak law of reductio ad absurdum $(A \to \neg A) \to \neg A$ is a special case of reductio ad absurdum, considering the law of identity $A \to A$, so from the premise $A \to \neg A$ follows $A \to A \wedge \neg A$, therefore $\neg A$.

In the particular case in which $\neg A$ replaces A everywhere, since $\neg\neg A$ is equivalent to A we have the *consequentia mirabilis:* $(\neg A \to A) \to A$, which so fascinated Gerolamo Saccheri (1667–1733).

Conversely, finally, admitting the reductio ad absurdum, modus tollens is justified because, assuming $A \to B$ and $\neg B$, from A with

$A \rightarrow B$ by modus ponens B follows, but we also have $A \rightarrow \neg B$ from $\neg B$ by the law of affirming the consequent; therefore also $\neg A$.

With the abundance of these equivalent inferential forms available for a variety of forms of discourse, we shouldn't be surprised by the frequency of negative arguments pointed out by Szabó.

On the other hand, the rule of modus ponens is

$$\frac{A \rightarrow B,\ A}{B}$$

with the chiastic structure:

$$
\begin{array}{l}
\lceil\ 1.\ \text{If A is true} \\
\ \ \lceil\ 2.\ \ \text{B is true.} \\
\ \ \lfloor\ 2^*.\ \ \text{A is true.} \\
\lfloor\ 1^*.\ \ \text{Then B is true}
\end{array}
$$

which is less interesting because the modus ponens doesn't need rhetorical devices to be grasped, it is wired into the neuronal circuits, even in animals.

In Euclid we also find a form that can be traced back to the ring composition but which isn't found in Greek rhetoric. Doxiadis finds such an absence plausible because it is a very elaborate construction and, in order for it to work, there are requirements to be satisfied that are not present in rhetoric's cultural context. Doxiadis calls it ring composition by (or with) substitution. An example is found in the proof for Proposition I.7 in the *Elements*, which states:

> Given two straight lines constructed from the ends A and B of a straight line and meeting in a point C, there cannot be constructed from the ends of the same straight line, and on the same side of it, two other straight lines equal to the given ones respectively and meeting in another point D.

Proposition I.7 is important for a number of reasons; it is essential for the following Proposition I.8, which states that if two triangles have the same base and also have the two corresponding sides equal, then the angles contained by the two sides will also be equal. But above all, as noted by Lewis Carroll (1832–1898), I.7 affirms the property of the rigidity of the triangle; according to some commentators it would also have been used in astronomy to demonstrate that three consecutive eclipses cannot occur at the same distance (see Heath's comments in Euclid 1956, 259–261).

The proof begins by assuming that there are two given straight lines *AD* and *BD* equal respectively to *AC* and *BC*, as in the diagram:

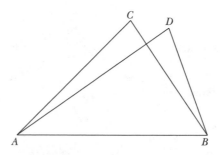

and it continues:

> Therefore, since *AC* is equal to *AD*, the angle *A$\hat{C}$D* is also equal to angle *A$\hat{D}$C* [Proposition I.5]. So *A$\hat{D}$C* is greater than *D$\hat{C}$B* [Common notion 5]. Thus *C$\hat{D}$B* is much greater than *D$\hat{C}$B* [Common notion 5]. Again, since *CB* is equal to *DB*, the angle *C$\hat{D}$B* is also equal to the angle *D$\hat{C}$B*. But it has been shown that the first angle is much greater than the second. This is impossible.

(The translation is literally faithful to the Greek text, but according to Lucio Russo we can read: "*a fortiori* greater than," where "much greater than" occurs; see Russo, Pirro, and Salciccia 2017, 51, n7.)

The proof therefore ends, as required, with the repetition of the statement, the sumperasma.

The primary difficulty of the proof is in the first step, demonstrating that the angle *A$\hat{D}$C* is greater than *D$\hat{C}$B*. There are three relevant relative statements.

1. Therefore, since AC is equal to AD, the angle $A\hat{C}D$ is also equal to angle $A\hat{D}C$.

2. So $A\hat{D}C$ is greater than $D\hat{C}B$.

3. Thus $C\hat{D}B$ is much larger than $D\hat{C}B$.

Although the three statements do not constitute a syllogism, there is a formal analogy due to the presence of a middle term, $A\hat{D}C$, which occurs in the first and second statements. If it were a syllogism, the conclusion of the third affirmation should connect the other two terms $A\hat{C}D$ and $D\hat{C}B$. But while $D\hat{C}B$ is mentioned, in the third statement $C\hat{D}B$ has taken the place of $A\hat{C}D$. To see where $C\hat{D}B$ comes from, it is necessary to fill the gaps in the argument. In a more complete version, the statements written in italics below ($1a$, $1b$, and $2a$) should be added.

1. Therefore, since AC is equal to AD, the angle $A\hat{C}D$ is also equal to angle $A\hat{D}C$.
   1a. Now $A\hat{C}D$ is greater than $D\hat{C}B$.
   1b. But $A\hat{C}D$ is equal to $A\hat{D}C$.

2. So $A\hat{D}C$ is greater than $D\hat{C}B$.
   2a. But $A\hat{D}C$ is a part of $C\hat{D}B$.

3. Thus $C\hat{D}B$ is much larger than $D\hat{C}B$.

Now we see that the two triples $1a$-$1b$-2 and 2-$2a$-3 respect the syllogistic form, with middle term and the connection between extremes.

If we merge the two triples using 2 only once, we get a perfect and symmetrical ring composition (the previous numbering is inserted in parentheses for comparison).

In the following arrangement, the pivots are shown in bold: 1-1^* pivots on $D\hat{C}B$, while 2-2^* pivots on $A\hat{C}D$.

> 1. (1*a*) Now $A\hat{C}D$ is greater than **$D\hat{C}B$**. [Common notion 5]
>> 2. (1*b*) But **$A\hat{C}D$** is equal to $A\hat{D}C$.
>>> 3. (2) Therefore $A\hat{D}C$ is greater than $D\hat{C}B$ [Common notion 5]
>> 2*. (2*a*) But **$A\hat{C}D$** is part of $C\hat{D}B$.
> 1*. (3) Therefore $C\hat{D}B$ is greater than **$D\hat{C}B$**. [Common notion 5]

This is the powerful inference that Doxiadis calls substitution RC; the substitution is that allowed by 1*b*. An analogy can be seen with the merging of two syllogisms into one, allowed by substitution, or even with the unification rule in modern resolution calculus.

The three basic forms that we have illustrated can be found combined in various ways in longer arguments, which are also structured in whole or in part as X or RC, and which Doxiadis calls "mesostructures."

The one which Doxiadis finds most aesthetically pleasing is the long RC, the only form of RC that is centered on a rule.

The long RC is illustrated by the proof of Proposition I.1. This proposition asks us to construct an equilateral triangle on a given base AB. As we know, the construction consists of tracing two circles with a radius $\overline{AB}$ with centers at A and B respectively, and to consider their intersection C.

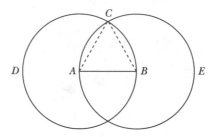

The existence of intersection C was contested to Euclid, in the absence of a principle of continuity, but this interesting question is not relevant here. The preceding drawing is the one found in the *Elements*.

In the presentation of the proof in RC form, the construction is summarized in the two combined statements 2 and 3 which describe the task, without going into detail. To understand the significance of the letters *C*, *D*, and *E* refer to Euclid's diagram.

In the ring composition, 2 and 2* pivot on the "given finite straight line"; 3 and 3* on "the triangle *ABC*"; 4 and 4* on "*CA*, *AB*, and *BC*"; 5 and 5* on the pair "*CA* and *CB*."

1. Construct an equilateral triangle on a given finite straight line.
   2–3. Let *AB* be the given finite straight line. Construct an equilateral triangle on *AB*.
      4. Since *A* is the center of the circle *CDB*, *CA* is equal to *AB*. And since *B* is the center of the circle *CAE*, *BC* is equal to *BA*. But *CA* was also shown to be equal to *AB*.
         5. Therefore *CA* and *CB* are both equal to *AB*.
            6. But things equal to the same thing are also equal to one another [Common notion 1]
         5*. Thus *CA* is also equal to *CB*.
      4*. Therefore the three straight lines *CA*, *AB*, and *BC* are equal to each other.
   3*–2*. *ABC* is equilateral, constructed on the given finite straight line *AB*.
1*. (Which is) what needed to be done.

In other cases, both the X and RC forms can be present in the same proof, possibly more than once, in a succession of substatements. Proposition I.2 reads:

Place a straight line equal to a given straight line at a given point (as an extremity).

The proof is based on the following drawing proposed by Euclid, otherwise it would be too difficult to follow.

After the statement, it begins with: Let *A* be the given point, and *BC* the given straight line; so it is required to place a straight line at point *A* equal to the given straight line *BC*.

The first part is simply the construction of what is shown in the drawing.

The straight line *AB* is drawn from *A* to *B* and the equilateral triangle *DAB* is constructed (Proposition I.1). The straight lines *DA* and *DB* are produced to *DE* and *DF* respectively (Postulate 2). The circle *CGH* is drawn with center *B* and radius *BC* (Postulate 3), and similarly the circle *GKL* with center *D* and radius *DG*.

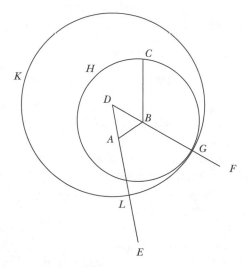

Then, the deductive part begins.

⌈ 1. Statement of the proposition.
 ⌈ 2. Point *A* and segment *BC* are given.
  ⌈ 3. And since *B* is the center of the circle *CGH*, *BC* is equal to *BG*.
   ⌈ 4. Again, since *D* is the center of the circle *GKL*, *DL* is equal to *DG*.
    [5. But *DA* is equal to *DB*.
   4*. Thus the remainder *AL* is equal to the remainder *BG*. [Common notion 3]
  3*. But *BC* was shown to be equal to *BG*.
  ⌈ 6. *AL* and *BC* are both equal to *BG*.
   ⌈ 7. But things equal to the same thing are also equal to one another. [Common notion 1]
  6*. Thus, *AL* is equal to *BC*.
 2*. Thus *AL*, equal to *BC*, was placed at point *A*.
 1*. (Which is) what needed to be done.

The reader can check that this is precisely the order of Euclid's text, without modifications. The first part is the ring composition 3-4-5-4*-3*; 3 and 3* pivot on the equality of BC and BG; 4 and 4* pivot on the lines DL and DG, from which the segments DA and DB are subtracted in 4*.

The priamel form has been mentioned in connection with rhetoric, with its poetic origins. In rhetoric, the counternarratives are discarded as red herrings in favor of that which is of most interest to the speaker. In mathematics these counternarratives take the form of counterstatements, that is, statements that contradict the theorem; these are then excluded as not true, rather than just "less probable" as in the legal context.

A good example is found in Proposition I.19, a portion of which we already examined in presenting a chiasmus. Let's take a look at the complete proof.

1. In any triangle, the greater angle is subtended by the greater side.
   2. If in the triangle ABC the angle $A\hat{B}C$ is greater than $B\hat{C}A$, then AC is greater than AB.
      3. If not, AC is either equal to or less than AB.
         4. In fact, AC is not equal to AB.
            5. Otherwise $A\hat{B}C$ would also be equal to $A\hat{C}B$.
            5*. But it isn't.
         4*. Therefore AC is not equal to AB.
         6. Nor, indeed, is AC less than AB.
            7. Otherwise $A\hat{B}C$ would also have been less than $A\hat{C}B$.
            7*. But it isn't.
         6*. Thus AC is not less than AB.
      3*. But it has been shown that AC is not equal to AB.
   2*. Therefore AC is greater than AB.
1*. Thus in a triangle, the greater angle is subtended by the greater side. QED

The two internal chiasmi 4-5-5*-4* and 6-7-7*-6* are triggered by 3. In what we might call a disjunctive syllogism iterated twice, each one could be developed in a ring composition, inserting a central statement that, in the first case, appeals to Proposition I.5, and in the second to Proposition I.18.

In 3, followed by two chiasmi, we recognize the origin of the inference which will come to be known as the disjunctive syllogism, that is to say the rule:

$$\frac{A \vee B, \neg A}{B}.$$

In the previous proof of I.19, we get to line 3 by imagining that we have inserted a statement "AC is greater than, equal to, or less than AB," and with a tacit disjunctive syllogism we have directly considered "if AC is not greater than, then it is equal to or less than AB." If inserted explicitly, we obtain:

1. In any triangle, the greater angle is subtended by the greater side.
   2. If in the triangle ABC the angle $A\hat{B}C$ is greater than $B\hat{C}A$, then AC is greater than AB.
      3. AC is greater than, equal to, or less than AB.
         4. If it is not greater than, AC is equal to or less than AB.
            5. In fact, AC is not equal to AB.
               6. Otherwise $A\hat{B}C$ would also be equal to $A\hat{C}B$.
               6*. But it isn't.
            5*. Therefore AC is not equal to AB.
            7. Nor, indeed, is AC less than AB.
               8. Otherwise $A\hat{B}C$ would also be less than $A\hat{C}B$.
               8*. But it isn't.
            7*. Thus AC is not less than AB.
         4*. But it has been shown that AC is not equal to AB and not less than AB.
      3*. Therefore AC is greater than AB.
   2*. Therefore if in the triangle ABC the angle $A\hat{B}C$ is greater than $B\hat{C}A$, then AC is greater than AB.
1*. Thus in a triangle, the greater angle is subtended by the greater side. QED

Doxiadis distinguishes this type of priamel, referring to it with a term sometimes used in logic: "exhaustion proof"; the simpler terms "proof by cases" or "proof by distinction of cases" are also found. We encounter this when particular restrictions are separately imposed on a statement in order to discuss their effects in greater detail or with greater ease, and to exclude all contingencies except one; the condition is that the various cases into which the problem is divided be exhaustive.

However, the purpose of this may be to deal with the cases separately rather than eliminate them. The first known mathematical example is the calculation of the area of lunes (squaring of lunes) done by Hippocrates of Chios around 430 BCE. Hippocrates' proof deals separately with three cases according to whether the outer boundary of the lune is a semicircle, greater than, or less than a semicircle. To focus on the context, consider the fact that this result was obtained during precisely the same years in which Gorgias and Antiphon were expressing themselves in the way we have documented above.

Finally, one should consider the case of the tree structures where elements of one X or RC branch out into other Xs or RCs, and again, with a nested iteration. However, there are rarely more than two levels. Curiously, although there are examples of these more complex structures in Homer and the lyric poets, in rhetoric—Gorgias, Antiphon, Lysias—the simple structures we have mentioned abound but not the complex ones. The earliest examples of these are found in Thucydides, primarily in parts where there is scant factual information. Scholars think that this is a kind of artificial "filler" which can be found in the written versions but that in the oral telling might have been too difficult to grasp (see Doxiadis 2012, 365; perhaps this is a good moment to mention that Doxiadis relies on various studies on rhetoric, which we have not cited as they are most likely only of interest to specialists).

To complete the framing of the X and RC rhetorical forms in classical Greek culture, it must be remembered that one of the most common mnemonic techniques was to associate the material to be remembered with particular locations in the mental representation of a journey through familiar places. X and RC are the natural forms for making explicit and developing particular deviations; the discursive style sometimes involves only the memory of a known result or law, other times it is mentioned explicitly; in these cases the demonstration is enriched with new RCs, which can be inserted, or not.

This expository policy is also typical of mathematical reasoning; it happens in Euclid that sometimes he repeats the result to which he refers, or a common notion or a postulate, word for word, other times not. An unsolved problem is understanding why Euclid explicitly cites a reference in some cases in the inartistic proof, for example repeating the statement of a common notion or that of a previous result, and other times not. On the other hand, in forensic rhetoric, it was customary not to repeat testimonies once they had been given (ibid., 338). Perhaps it is impossible to understand Euclid's choices, not knowing the purpose of the *Elements*, their use, and the intended audience (ibid., 365–369).

Conditioned by Aristotle, one thinks that logic is expressed through chains of perfect syllogisms, but in the case of rhetoric, the inferences are primarily enthymemes, that is, syllogisms flawed in one of two senses: they have premises that are probable rather than certain, or some premises are missing. In this latter sense, the style of Euclid and of Greek mathematics in general, like that of today, is strongly enthymematic.

7.2 *Data*

We end with an example taken from Euclid's *Data* to show that the Greeks considered presentation in the form of X or RC as a sine qua non.

The *Data* (from δέδομαι, δεδομένος, Attic, otherwise δίδωμι) are, according to Pappus, who lived in Alexandria during the reign of Diocletian (284–305), "the first of the preparatory works for the acquisition of skills suitable for solving geometric problems with the technique of analysis" (see F. Acerbi's introduction in Euclid 2014, 439–554, esp. 519–523).

Basically *Data* is a collection of constructions; it is clear that a systematic catalog of constructions would have facilitated the analysis procedure, which is why Pappus inserted the text in the *Treasury of Analysis*, in book VII of his *Synagoge* or "collection" of mathematical material. The purpose of *Data* is to present a series of theorems collected together that are available for any use and are structured as data chains. The book begins with the definitions of the senses in which objects are said to be given, the first being: "Given in magnitude is said of figures and lines and angles for which we can provide equals." Data are objects assigned by hypothesis or obtainable from the starting data by means of geometric constructions or previous theorems. The treatise "[deals with] the transferability of the predicate 'given' by certain objects to others" (Euclid 2014, 522).

The data can be specified in magnitude, in position, or in shape according to the geometric object under consideration; points, lines, and angles are given in position "if they always occupy the same place"; a circle is given in position and in magnitude when the center is given in position and the radius in magnitude, or only in magnitude when only the radius is given in magnitude. But ratios can also be objects, and ratios are given in absolute terms. An essential constraint is that the inferences with which the data are obtained must be expressed in the language of the "givens."

It is to be expected that many arguments that appear in the *Elements* will also appear in the *Data* in the version characteristic of this book. If a proposition from the *Elements* solves a problem by giving us instructions to find an object, the proof proves that the object is given. The other type of proofs that we find in the *Elements* comprises those that establish a property of a given object, often in the first part by proving that it is given.

The English translator of *Data*, Christian M. Taisbak, admitted in the introduction to his translation that he was amazed by the constant repetition of certain words, like "if this item [object, proposition] is given, then this other is also given," to the extent that in an early version he decided to remove many of those repetitions, and words like "also." Then as he delved further into the work he discovered that he was leaving out an essential feature of the exposition, which aimed to present the data linked together in chains and every proposition needed to establish new links between them.

We give two examples, based on Taisbak's 2003 translation of Euclid but the original Greek letters have been used. (Doxiadis's text on this subject is confusing, perhaps due to a typo, because it gives the protasis of *Data* 4, but the proof of *Data* 5, and this is the demonstration to which he wants to draw attention. For the convenience of the reader we present both propositions. *Data* 4 is simple and presents no difficulties.)

Data 4:   "If a given magnitude be subtracted from a given magnitude, the remainder will be given."

The accompanying figures are:

and

Proof: "From a given magnitude AB, let a given magnitude $A\Gamma$ be subtracted. I say that the remainder ΓB is given.

For, since AB is given it is possible to provide a magnitude equal to it. Let it be produced and be ΔZ. Again, since $A\Gamma$ is given it is possible to provide a magnitude equal to it. Let it be produced and be ΔE. Then, since AB is equal to ΔZ, and $A\Gamma$ to ΔE, the remainder ΓB is equal to the remainder EZ: Therefore ΓB is given; for EZ has been produced equal to it."

As an exercise, the reader can try to transcribe the proof into the RC form.

Data 5: "If a magnitude have a given ratio to some part of itself, it will also have a given ratio to the remainder."

The accompanying figures are:

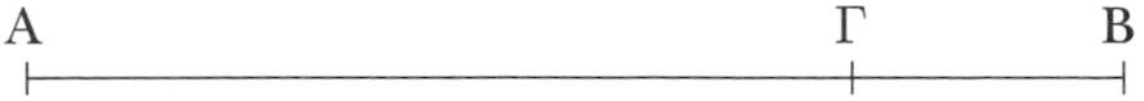

and

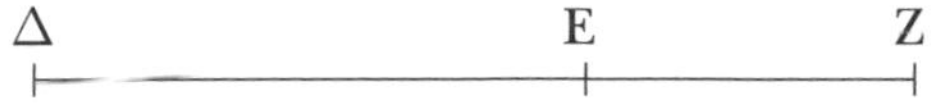

Proof: "Let a magnitude AB have a given ratio to some part of itself $A\Gamma$. I say that it has also a given ratio to the remainder $B\Gamma$.

For, let a given magnitude ΔZ have been set out. And since the ratio of $BA{:}A\Gamma$ is given, let $Z\Delta{:}\Delta E$ have been provided the same as it. Then the ratio $Z\Delta{:}\Delta E$ is given. And $Z\Delta$ is given. Therefore ΔE is given. Therefore the remainder EZ is given. And ΔZ is given: therefore the ratio $\Delta Z{:}ZE$ is given. And since $\Delta Z{:}\Delta E :: AB{:}A\Gamma$ [ΔZ is to ΔE as AB is to $A\Gamma$], thus by *convertendo* [turning it around] [$\Delta Z{:}(\Delta Z-\Delta E) :: AB{:}(AB-A\Gamma)$, that is] $\Delta Z{:}ZE :: AB{:}B\Gamma$. And the ratio of $\Delta Z{:}ZE$ is given, as has been proved: therefore the ratio of $AB{:}B\Gamma$ is also given."

The proof certainly appears convoluted and difficult to follow; it is easier to read if we put it into the form of a ring composition, which is really the form that it naturally follows (according to Doxiadis 2012, 371–372).

We'll use the abbreviation d(x) to mean "x is given," and $X{:}Y$ means "X is to Y," "ratio of X to Y."

The proof can be presented in this form of RC, where the pivots are in bold:

⌈ Let d(AB:AΓ). I say that **d(AB:BΓ)**. Construct ZΔ.
　⌈ Since d(**AB:BΓ**), construct **ZΔ:ΔE** equal to it. (*Data* 2)
　　⌈ So **d(ZΔ:ΔE)**.
　　　⌈ And **d(ZΔ)**.
　　　　[Therefore d(EZ). [EZ is the remainder.] (*Data* 4)
　　　⌊ Thus **d(ZΔ)**.
　　⌊ But also **d(ZΔ:ΔE)**.
　⌊ Since AB:AΓ :: ΔZ:ΔE, turning it around, **ΔZ:ZE :: AB:BΓ**.
⌊ Therefore also **d(AB:BΓ)**.

As we have seen, the examination of the work of Euclid—the inventor, or at least the oldest author whose proofs have survived, and which have characterized the history of Western mathematics—highlights the range of cognitive potential offered by narrative and the cultural tools of poetry. Having fully considered it, it doesn't seem so absurd to think that the proof is distantly derived from literature.

Conclusions

> Our existence is a pale ghost; we live it, but actually only on the basis of an error.
> —Robert Musil

Even those, like R. D. Carmichael (1879–1967), who believe that mathematics and poetry are thought's polar opposites, respectively systematic and nonsystematic, admit that the products of both disciplines have a common quality—millennial permanence. Carmichael marvels at a historical coincidence: "It is a matter of great inspiration to see the Greek geometry and the Greek tragedy surviving through the ages and retaining the active power to excite our admiration and increase our happiness today" (Carmichael 1930, 244). He doesn't say so specifically, but the paring of geometry and tragedy cannot but be suggested by the contemporaneity, as well as by the fact that the two phenomena occurred in the same place. If Carmichael had had the tools, developed later, to further explore his observation, perhaps he would have been induced to change his opinion.

Affinities between poetry and mathematics, beyond their common ancient origins, are a recurring theme among contemporaries, even among professional mathematicians. Marguerite Lehr (1898–1987) highlights a lesson derived from poetry, that is, the need of a metaphorical language, even for mathematics: "Mathematics and poetry are head-tail of the same coin. Both are an attempt to discern pattern—because you refuse the idea of a chaotic universe—and to develop language which has to be metaphoric, or you'll drown in jargon" (Kenschaft 1988, 9). Jargon can be used to facilitate communication in homogeneous groups, but also to hide and to exclude. Given the context of the phrase, in the case of mathematics we can understand "jargon" not only as "technical language," but also an "incomprehensible language." A metaphorical language can only be a language of stories.

For others, the common element is conciseness:

> Voltaire remarked that "one merit of poetry few persons will deny; it says more in fewer words than prose"; and why may we not with perfect truth continue, "one merit of mathematics no one can deny—it says more in fewer words than any other science in the world"? (David Eugene Smith (1860–1944), quoted by Schmalz 1993, 132).

Similarly, Lipman Bers (1914–1993) thought that "mathematics is very much like poetry. I think that what makes a good poem—a great poem—is that there is a large amount of thought expressed in very few words. In this sense formulas are like poems." Bers, fascinated by Pindar's poetry, introduced to him by his son, asked him if a good translation existed. His son's reply was: "There is no good translation. Either you read it in Greek, or you just forget it!" Convinced, Bers added, "As far as mathematics is concerned, something similar may be true. Either you can use symbols, or just forget it!" (Albers and Reid 1987). The challenge of expression through symbols was discussed in chapter 5. The quality of conciseness, or compression, is discussed in Lolli (2016b).

At least the beauty of mathematics, always elusive and difficult to define, and even more so the beauty of mathematical reasoning, is decisively confirmed if, at its origin—the origin of deductive mathematics—we can find expressive forms with the cadence, the rhythm, the ability to capture a person's attention and be followed like the pied piper, the same as those which radiate from verses organized by X and RC, and were tried in poetry: the beauty that enchains.

A few references on the theme of beauty will suffice, yet not unknown ones. A famous statement by Hermann Weyl (1885–1955) recalled by Freeman Dyson (1923–) says: "My work always tried to unite the true with the beautiful; but when I had to choose one or the other, I usually chose the beautiful." The physicist Subrahman-yan Chandrasekhar (1910–1995) asked Dyson to shed more light on the episode and he reconstructed it, recalling that Weyl cited an example in which he had sacrificed truth: his gauge theory of gravitation, which he had worked out in his *Raum-Zeit-Materien*. Apparently, Weyl had doubts about the theory but didn't abandon it for the sake of its beauty. Later, it did turn out that Weyl's instinct was right and his formalism of gauge invariance was incorporated into quantum electrodynamics, yet another example of seemingly gratuitous mathematical creations that turn out to be useful later in unexpected ways (see Schmalz 1993, 140).

Godfrey H. Hardy devoted a number of scattered remarks to the concept of beauty in mathematics. He maintained that both generality and depth are features of mathematical beauty. This had not yet been defined, but Hardy was sure that no one would dispute his examples, which are Euclid's theorem on the infinity of prime numbers and that of Hippasus of Metapontum (ca. 500 BCE), a Pythagorean, on the irrationality of $\sqrt{2}$ (Hardy 1967, 92–94 and 94). The aesthetic criteria to which Hardy appeals to affirm the beauty of these theorems are unexpectedness, inevitability, and economy of the methods of proof (ibid., 113).

In his *Apology*, Hardy also defined the mathematician as a builder of patterns, like the painter or the poet, and forged the definitive statement: "Beauty is the first test: there is no permanent place in the world for ugly mathematics" (ibid., 85).

However, it is not a given that the goal of the poet is beauty, unless indirectly, as an added value. It is also incorrect to think of the poet as sentimental, as David Eugene Smith imagines when he states: "What, after all, is mathematics but the poetry of the mind, and what is poetry but the mathematics of the heart?" Sofya Kovalevskaya (1850–1891), a mathematician at a time when it was difficult for a woman to practice that profession, and a courageous writer, has a penetrating look at both the poet and the mathematician, observing: "it is impossible to be a mathematician without being a poet in soul." But to understand it

> one must renounce the ancient prejudice that a poet must invent something which does not exist, that imagination and invention are identical. It seems to me that the poet has only to perceive that which others do not perceive, to look deeper than others look. And the mathematician must do the same thing. (Osen 1974, 136)

She must, but the fact is that mathematicians can, because, thanks to how they have trained their imaginations, or because of their DNA, it's as if they have a sixth sense. Darwin acknowledged this in his autobiography:

> and in after years [following his studies at Cambridge] I have deeply regretted that I did not proceed far enough at least to understand something of the great leading principles of mathematics; for men thus endowed seem to have an extra sense. (Darwin 1958, 58)

We don't know if he is alluding to a specific example (and if so which) but certainly he would have felt the same enthusiasm that Hilbert felt when he talked about the *Drosophila*, noting that *Drosophila* is a small fly, but our interest in it is great. The reason was that he saw in the hereditary laws of the *Drosophila* a similarity with the application of the axioms of linear congruence, "so simple and exact, and at the same time so wonderful, that probably no fantasy, no matter how bold, could have devised it" (Hilbert 1930; English translation in Ewald 1996, 1159).

Darwin's method was careful, patient, loving observation. The same could be said of the ancients observing the sky; they were prudent in constructing generalizations and hypotheses, and made few errors. Darwin summed up the qualities that made him a successful scientist, apart from his obvious love of science itself, as "unbounded patience in long reflecting over any subject—industry in observing and collecting facts—and a fair share of invention as well as of common-sense" (Darwin 1958, 145). He almost always had to abandon or modify the first formulation of a hypothesis, because he always endeavored to keep his mind "free, so as to give up any hypothesis, however much beloved." Some of his critics said that he was a good observer but with little capacity for reasoning; Darwin rejected this accusation, remarking: "I do not think that this can be true, for the *Origin of Species* is one long argument from the beginning to the end, and it has convinced not a few able men" (ibid., 140–141).

Toward the end of his autobiography, he admitted, "my mind has changed during the last twenty or thirty years"; he had almost completely lost his taste for painting and music, he found Shakespeare unbearably heavy. Yet up until his thirties he had taken great pleasure in reading the work of a wide variety of poets, and had loved Shakespeare very much since his boyhood.

> This curious and lamentable loss of the higher aesthetic tastes is all the odder. . . . My mind seems to have become a kind of machine for grinding general laws out of large collections of facts, but why this should have caused the atrophy of that part of the brain alone, on which the higher tastes depend, I cannot conceive. (Ibid., 139)

We understand it better, and we have serious reasons for concern, now that we are faced with "big science" and, in many fields, work almost exclusively focused on analysis of "big data." Our concerns are not about the possibility of a decline in an individual's love of the arts, but about the ongoing changes in science. Scientists are now becoming experts who interpret the natural world and who are given the task of transforming knowledge into power and profit. Their moral qualities are perhaps no different from those of their predecessors (as Steven Shapin (1943–) maintains, in Shapin 2008;

see also Pacchioni 2017). However, the planning of truth production, and the depersonalization of the instruments of scientific output, seem to be leading toward the elimination of the personal
aspects of the equation in science and industry. It seems that scientists are primarily the administrators of big budgets (Weinberg
1967). There is still hope in pure mathematics.

Albert Einstein (1879–1955) didn't have any data to put into the
equations when he understood that gravity causes space to curve
(Frenkel 2013, 2). There were no such data. No one could even
imagine that space is curved, and what that means. At most, we can
visualize surfaces (two-dimensional spaces) in space. The difficulty
was even greater than that encountered by Galileo Galilei in supporting the fact of the earth's diurnal and annual motion.

In the case of the earth's diurnal motion, visual experience confirmed common-sense conclusions: the cannonball falling from the
ship's mast, the shot fired toward the east and the west, the birds
circling in place:

> for just as I . . . have never seen nor ever expect to see the rock fall
> any way but perpendicularly, just so do I believe that it appears to the
> eyes of everyone else. It is therefore better to put aside the appearance,
> on which we all agree, and to use the power of reason either to confirm
> its reality or to reveal its fallacy. (Galilei 1967, 256, second day)

The dialogue proved that common experiences were compatible
with both hypotheses, either stationary or diurnal motion. Galileo's discourse was made of arguments and analogies, stories, like
that of an incident from which

> one may learn how easily anyone may be deceived by simple appear-
> ances, or let us say by the impressions of one's senses. This event is
> the appearance to those who travel along a street by night of being
> followed by the moon, with steps equal to theirs, when they see it
> go gliding along the eaves of the roofs. There it looks to them just as
> would a cat really running along the tiles and putting them behind
> it; an appearance which, if reason did not intervene, would only too
> obviously deceive the senses.

The case of the annual movement was more challenging, such that he could not "sufficiently admire the outstanding acumen of those who have taken hold of this opinion and accepted it as true; they have through sheer force of intellect done such violence to their own senses as to prefer what reason told them over that which sensible experience plainly showed them to the contrary" (ibid., 328, third day). He then expresses his admiration a second time: "there is no limit to my astonishment when I reflect that Aristarchus and Copernicus were able to make reason so conquer sense."

Criticism of passive observations and the choice of other, mathematically driven observations were the plot of Galileo's dialogue, but Einstein had no observations to offer or criticize. The curvature of space did not manifest its effects in any everyday experience. Of necessity he had to listen only to what the discourse dictated to him. However, Einstein's imagination was supported by powerful mathematical concepts that allowed him to conceive spaces free from the constraints of the Euclidean one. To have a geometry based on the varieties introduced by Bernhard Riemann (1826–1866), it suffices that a distance function is defined for each pair of points. So that Einstein was able to conclude his discourse and, as the Italian poet Manzoni said of Napoleon, "gain the goal and grasp the prize, that t'was folly to aspire." This explains how he, the physicist, came to the conclusion that "the creative principle [of science] resides in mathematics" (Einstein 2011, 385). We also understand why scientists like Frenkel (2013) speak of their research as an exploration into "the hidden heart of reality," not unlike the exploration of nature's secrets in the sixteenth century, which to the layman look like fantasies of delusional universes.

Perhaps it is only in mathematics that we encounter the symbiosis of necessity with Hamlet's dilemmas, which was the characteristic of the archaic myth. Hilbert gave voice to widespread and inevitable sentiment when he stated:

Mathematics is in no sense like a game, in which certain tasks are determined by arbitrarily established rules. Rather, it is a conceptual system guided by internal necessity, that can only be so, and never otherwise. (Hilbert 1992, 14)

And yet the mathematicians have once again found a way to overcome Fate, as de Santillana said, by "making of Necessity a free creation" (quoted in section 1.1). It turns out that Necessity "encompasses some of the most entertaining and intense adventures of human existence" (Musil 1990, 41). Mathematics, after Riemann, Poincaré, and Hilbert himself, started to fly high, like the island of Laputa (the flying island inhabited by scientists in *Gulliver's Travels*). Now the mathematicians "actually looked all the way to the bottom and found that the whole building was standing in mid-air" (ibid., 42). Musil is referring to the so-called crisis in the foundations of mathematics, in full swing at the time he was writing. But since the foundations were not found, and don't exist, his observations are still valid today, and the building is standing in midair. We live our lives on the basis of an error, Musil says, or on the basis of a fantastic tale.

Appendix

Notation and terminology

1. In the original Italian text the word *senso* was used where in English one would use "meaning" which is usually rendered *significato* in Italian. In Italian, *senso* and *significato* are often used interchangeably, with different nuances in some expressions. However, these words have an important place in analytical philosophy, starting from *Über Sinn und Bedeutung* (1892) by Gottlob Frege (1848–1925). In this essay, Frege says that the *Bedeutung* (referent) of a sign is the object it designates, whereas the *Sinn* (sense) is the manner in which it is designated. In Italian, the title of Frege's essay is translated as *Senso e significato*, with the meaning (forgive the use of that word) *Senso e denotazione* (Meaning and denotation). In English it has been translated as *Sense and Reference* in Blackwell's 1952 edition; but later it was also translated as *Sense and Meaning*. In his 1905 *On Denoting*, Bertrand Russell affirmed that Frege distinguishes between meaning and denotation. In the Italian text, I have used the word *senso* (meaning) to translate the German *Sinn* and *denotazione* for *Bedeutung*. In order to avoid confusion, the word *denotazione* is only used when Frege's *Bedeutung* is intended.

[Translator's note: While I have endeavored to follow the author's intent in my translation from the Italian, "denotation" in English has not proved the appropriate translation for *significato* in Italian in every context.]

2. Rudolf Carnap (1932) developed a physicalist conception of language, in which there were no longer autopsychic phenomena (*eigenpsychische Phänomene*), as in his previous *Der logische Aufbau der Welt* (1928), but intersubjectively verifiable physical objects that must form the fundamental reference. Carnap's physicalist or physical language is the language used in physics. This includes so-called "thing language," which has common nouns for objects and "thing" for general statements, in addition to the scientific terminology necessary for a scientific description of inorganic processes. For example, he notes that "soluble" and "elastic" are not observable predicates and therefore are not part of thing language; they are theoretical terms (Carnap 1936, 466). Carnap later liberalized these constraints, also allowing introspection and inference by analogy. Otto Neurath (1882–1945) attempted a physicalist reduction of the social sciences.

3. The text refers frequently to Euclid's *Elements* (Στοιχεῖα). Remember that the thirteen books of the *Elements* deal not only with geometry but also with number theory and the theory of proportions. The propositions are indicated with the Roman numeral of the book and the Arabic numeral of the proposition.

4. The symbol N is used to indicate the set of natural numbers $\{0, 1, 2, \ldots, n, \ldots\}$; and $\mathbb{R}$ the set of real numbers. $\mathbb{R}^n$ is the Cartesian product with n factors of $\mathbb{R}$, i.e., the set of n-tuples of real numbers with the operations pointwise defined on the coordinates. $\mathbb{C}$ is the set of all complex numbers.

In mathematics, by "continuum" we mean a completely ordered set, dense and such that each upper-bounded set has a least upper bound. According to the definition, the set of real numbers $\mathbb{R}$ is a continuum, as are multidimensional spaces such as $\mathbb{R}^n$. There are also topological definitions, such as Cantor's which defined a continuum as a perfect, connected set (Cantor 1883), and others. We also use the word "continuum" to talk about straight lines and other geometric figures in space, but for these we need an axiom of continuity, otherwise the straight line could only contain images of algebraic numbers.

Continued fractions are expressions of the form

$$x = a_1 + \cfrac{1}{a_2 + \cfrac{1}{a_3 + \cfrac{1}{a_4 + \ldots}}}$$

where $(a_1, a_2, a_3, \ldots)$ is called the development of the fraction, or of x, and the a_i are positive integers. If x is rational, its development is finite and there are at least two of them; if it is irrational, it is infinite and unique.

5. $\{x, y\}$ is the set (the unordered pair) that has elements x and y. The union $\cup\, x$ is defined as the set that contains the elements of the elements of x. The ordered pair $\langle x, y \rangle$ is the set $\{\{x\}, \{x, y\}\}$.

The set with three distinct elements x, y, z is $\{x, y, z\} = \cup\{\{x, y\}, \{z\}\}$, and similarly in general those with n elements. Hereditarily finite sets are finite sets, of finite sets, of finite sets ... down to $\emptyset$.

The Cartesian product $X \times Y$ is the set of ordered pairs $\langle x, y \rangle$ such that $x \in X$ and $y \in Y$. $X \subseteq Y$ means that X is a subset of Y.

A relation is a set of ordered pairs. A binary relation R between X and Y is a relation $R \subseteq X \times Y$. The domain of R, $dom(R)$, is the set of $x \in X$ for which there exists a $y \in Y$ such that $\langle x, y \rangle \in R$. The image of R, $im(R)$, is the set of $y \in Y$ for which there exists an $x \in X$ such that $\langle x, y \rangle \in R$.

A function between X and Y is a relation R between X and Y such that for every $x \in dom(R)$ there is only one $y \in Y$ for which $\langle x, y \rangle \in R$; this y is indicated by $R(x)$.

A function of f from X into Y, $f : X \rightarrow Y$, is a function between X and Y with $dom(f) = X$.

A one-to-one correspondence, or a bijective one or a bijection as it's also now called, is a function f of a set u in a set v, $f : u \rightarrow v$

injective and surjective; a function f is called injective if to the different elements of the domain u it makes correspond as values different elements of v; it is called surjective if every element of v corresponds with any element of u, that is, if the image of f is all v. An injective function f allows an inverse, f^{-1}, from its image to its domain, for which $f^{-1}(f(x)) = x$ is valid.

A sequence of elements of A is a function $f: \mathbb{N} \to A$, a sequence with repetitions if f is not injective. An enumeration of a set A is a function $f: \mathbb{N} \to A$, injective and surjective, or a one-to-one correspondence with $\mathbb{N}$. An effective enumeration is an enumeration that is a calculable function.

A set is countable, or of cardinality $\aleph_0$, if there is a one-to-one correspondence between the set and $\mathbb{N}$. Initially, the cardinality of a set was also called "power," but this term was abandoned because it is ambiguous in that it is also used for $\mathcal{P}(x)$, the set of all subsets of x. Two sets with the same cardinality, in the sense that there is a bijection between them, are called equipotent.

Finite ordinals, or natural numbers, can be defined in more than one way. The von Neumann numbers are $\emptyset$, $\{\emptyset\}$, $\{\emptyset, \{\emptyset\}\}$, $\{\emptyset, \{\emptyset\}, \{\emptyset, \{\emptyset\}\}\}$, . . . ; the Zermelo numbers are $\emptyset$, $\{\emptyset\}$, $\{\{\emptyset\}\}$, $\{\{\{\emptyset\}\}\}$, . . .

A total order of a set A is a relation $< \subseteq A^2 = A \times A$ which is transitive ($x < y$ and $y < z$ implies $x < z$), antireflexive ($a \not< a$) and connex (for every x, y, $x \neq y$, either $x < y$ or $y < x$). A well-order of a set A is a total order of A such that every nonempty subset $B \subseteq A$ has a $<$-minimum.

In ZF theory with the axiom of foundation, the ordinals are defined as transitive sets, well-ordered by the membership relation $\in$. A transitive set x is a set such that the elements of its elements belong to x.

After the rigorous development of the theory of ordinals, the cardinal numbers are defined as initial ordinals, that is, ordinals that do not have a one-to-one correspondence with any lesser ordinal; previously it was thought that they were classes of equivalence with respect to the relationship of being in one-to-one correspondence; however, these classes are too large to be sets. A difference in formulation or style remains: a result on cardinals should now only be proven on the basis of the properties of the correspondences (this is not always possible without complications).

It is shown that in ZF, with the axiom of foundation, there is a theorem to the effect that the universe is made as described in section 1.2. The levels are obtained with the recursive definition $V_{\alpha+1} = \mathcal{P}(V_\alpha)$ and $V_\lambda = \cup_{\beta < \lambda} V_\beta$, if λ is a limit ordinal.

The axiom of choice states that for every nonempty set A of nonempty sets there is a choice set: that is, a set such that its intersection with each element of A has only one element. In the case of finite sets, the choice set can be constructed; generally, for infinite sets it cannot be defined by a law.

6. Some geometries are described in section 1.3. The cross-ratio, or anharmonic ratio, of four aligned points A, B, C, D is the ratio $\dfrac{AC \cdot BD}{BC \cdot AD}$, where AC, AD, BC, BD are the lengths of the oriented segments. The cross-ratio doesn't change if the four points are projected onto another line with central projection.

7. A group is a structure with an associative operation, a neutral element, and the opposite of each element. If the operation is commutative, it is called an abelian group.

"Field" is a technical word that indicates an algebraic structure with operations like those of real numbers: with respect to addition, it is a commutative group; multiplication satisfies the associative and commutative properties and the existence of the neutral element and of the inverse for nonzero elements; and the distributive property of multiplication over addition is valid.

The concept of field extension is simple, apart from the details. For example, to add the solutions of $x^2 - 2 = 0$ to the rational numbers, we consider the set of terms $x \pm y\sqrt{2}$ where x, y are rational; they form a field; the correspondence $\sqrt{2} \to -\sqrt{2}$ is a symmetry of the extended field.

A Galois group is a group associated with a field extension. If A is an extension of B, a B-automorphism is an automorphism of A (a bijective correspondence of A to itself which preserves the operations and the neutral elements) and leaves the elements of B fixed. The B-automorphisms of a field A form a group, called a Galois group.

Finite fields modulo p contain the elements $\{0, 1, \ldots, p-1\}$ with the modulo p operations of addition and multiplication: which means the usual operations are performed, the result is divided by p, and the remainder, which is ≥ 0 and $\leq p - 1$, is taken as a result.

Harmonic analysis studies the representation of functions as the superposition of waves.

Automorphic functions are functions with complex values, introduced by Henri Poincaré (1854–1912), which are invariant under rational linear transformations and generalize trigonometric functions.

A Riemann surface, or a complex manifold of dimension 1, is a topological surface on which a complex structure is defined; a topological surface is a connected topological Hausdorff space M, with a countable base and such that every point $a \in M$ has an open neighborhood homeomorphic to an open set of $\mathbb{C}$ (homeomorphic means in one-to-one and bicontinuous correspondence); a complex structure on a topological surface M is given by an equivalence class of complex atlases; each complex atlas is a collection of charts, where each chart is a homeomorphism between an open set of M and one of $\mathbb{C}$ and the open sets form a covering of M (with compatibility conditions that we do not specify).

A torus is a donut-shaped surface.

A vector space is a structure composed of a field of scalars and of a set with elements called vectors, with two binary operations, addition and multiplication by a scalar.

8. A category $\mathcal{C}$ is formed of "objects" and "morphisms," indicated by $u : a \to b$ where a and b are objects, and there is an associative composition operation $\circ$ of morphisms, and for every a an

identity morphism $i_a : a \to a$. From the axiomatic perspective, category theory is identified by the axioms just set out.

The different categories can be distinguished according to the special objects they possess; for example, an object 1, called terminal, such that for every a there exists a single morphism $a \to 1$. The corresponding set-theoretic term, if the morphisms are thought of as functions, could be $\{\emptyset\}$, or any singleton.

Intuitively, the objects of a category could be structures of the same type, for example groups, and the morphisms the homomorphisms between the structures of that type, that is, the functions that preserve structure. However, a category is itself a structure, and therefore morphisms can be considered among categories, which are called functors.

A functor between two categories C and D is a pair of functions $\langle F, G \rangle$, in which F associates with every object of C an object of D and G associates with every morphism of $f : A \to B$ of C a morphism $G(f) : F(A) \to F(B)$ of D (if the functor is covariant, otherwise $G(f) : F(B) \to F(A)$ if the functor is contravariant), such that $G(i_A) = i_{F(A)}$ and $G(f \circ g) = G(f) \circ G(g)$ (or for contravariant functors $G(f \circ g) = G(g) \circ G(f)$).

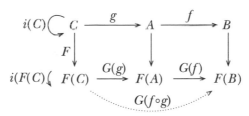

The conditions for a covariant functor are summarized by the commutative diagram in the figure above (note that the vertical arrows are not morphisms); a diagram is commutative if all of the paths (which are compositions of arrows) with the same origin and the same target are equivalent (in the figure, $G(f) \circ G(g) = G(f \circ g)$).

The lower morphism is punctuated to express the fact that its existence is affirmed, given $G(g)$ and $G(f)$, to make the lower diagram commutative. Category theorists have developed diagrams that express a rich schematic language.

With functors as morphisms and categories as objects we obtain a category *Cat*; *Cat* is the category of all categories, an entity which obviously contradicts the distinction of logical types.

An example of the usefulness of these diagrams is given by the categorical characterization of natural numbers. In a category with suitable closing properties, in particular that has a terminal object 1, the object "natural numbers" is presented as a pair of morphisms involving an object N

$$1 \xrightarrow{o} \mathbb{N} \xrightarrow{s} \mathbb{R}$$

such that for each pair $1 \xrightarrow{x} A \xrightarrow{f} A$ there is a unique h for which

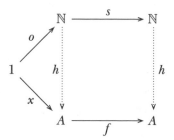

the diagram is commutative (when speaking, it's not even necessary to say: "the diagram is commutative"; we get to "therefore" or "such that" and we show the diagram itself; when it's drawn, the diagram is commutative).

The laws of category theory are almost always expressed with a commutative diagram.

In the diagram for the object "natural numbers" there appear those that in intuitive terms are 0, identified by $1 \xrightarrow{o} \mathbb{N}$ as an image of ∅ by means of o, if $1 = \{\emptyset\}$, and the successor function $\mathbb{N} \xrightarrow{s} \mathbb{N}$.

The diagram condenses two substantial properties of natural numbers: one is the minimality of N between simply infinite systems (according to Dedekind's definition of an infinite system: endowed with an injective function in itself that is not surjective); the other is the principle of definitions by recursion, also proved by Dedekind (for primitive recursive functions).

These two properties are immediately equivalent in category-theory language due to the simple fact that they are expressed by the same diagram as shown above. If (returning to the language of set theory, even with categorical notation) A is an infinite set with injection f in itself and $a \notin im(f)$, then if $1 \xrightarrow{x} A$ identifies a, h establishes the immersion of $\mathbb{N}$ in A. On the other hand, if you want to define a function h with the equations

$$\begin{cases} h(0) = a \\ h(s(x)) = f(h(x)), \end{cases}$$

the diagram states that in correspondence with a and f there exists the desired $h : \mathbb{N} \to A$ that satisfies the two equations.

9. A sheaf consists of the assignment of an $F(U)$—which can be a set, a structure, a vector space, a set of functions or something else—to each open subset of a topological space X. First, we define a presheaf: if $\mathcal{C}$ is a category, a presheaf associates an object $F(U)$ to every open U of X, and to every open pair U, V, with $V \subseteq U$ a morphism $res_{V,U} : F(U) \to F(V)$ in $\mathcal{C}$. The res restriction morphisms must satisfy: $res_{U,U} = i_{F(U)}$ and $res_{W,V} \circ res_{V,U} = res_{W,U}$ for $W \subseteq V \subseteq U$. $F(U)$ is called the section of F over U.

A sheaf is a $\mathcal{C}$-valued presheaf on X that satisfies two conditions: (1) $F(\emptyset)$ is the terminal object in $\mathcal{C}$, which is the object for which there exists exactly one morphism from every other object; (2) a gluing axiom, in two parts: we suppose that $U = \cup_i \in _I U_i$; if s, $t \in F(U)$ are such that for every $i \in I$ $res_{Ui,U}(s) = res_{Ui,U}(t)$, then $s = t$; whenever we choose elements $s_i \in F(U_i)$ in such a way that for every $i, j \in I$ we have $res_{Ui \cap Uj,Ui}(s_i) = res_{Ui \cap Uj,Ui}(s_j)$, there exists $s \in F(U)$ such that $res_{Ui,U}(s) = s_i$.

The gluing axiom allows you to move from a collection of local data, for each open set, to a datum defined over all of the topological space.

The terminology itself suggests that the most natural example of a sheaf is the sheaf of real-value functions over a topological space. 10. The language of first-order logic is an infinite list of individual variables, and symbols for relationships (at least one), for functions, and for special elements. Logical symbols are a set of propositional connectives, usually ∧, ∨, ¬, →, ↔ ("and," "or," "not," "if then," "if and only if") but also any other adequate system to express all truth functions, such as → and ¬, or ∨ and ¬; the quantifiers ∀ and ∃; and brackets.

In every interpretation, given by a structure, the variables take values (that vary) in the elements of the structure. A formula is logically valid if it is valid in every interpretation, that is, satisfied by every value of its free variables.

Propositional logic is the branch of logic without individual variables and quantifiers and with a list of propositional variables that, in every interpretation, take on values of 0 or 1. The formulas are then called propositions. Tautologies are propositions that have value 1 in any interpretation. "Law" is another way of saying "tautology."

The language of second-order logic also includes an infinite list of second-order variables and the symbol ∈. In each interpretation, the second-order variables take values in the domain's subsets. If the interpretations are defined so that the second-order variables vary on all subsets of the domain, the corresponding logic is called full.

A logical calculus is a set of formulas (logical axioms) and rules, by means of which the concept of derivation is introduced, that is, a finite succession (or a tree) of formulas in which each one is either obtained from precedents with a rule, or is one of the axioms.

Just remember that in natural-deduction calculus there are no logical axioms and there are introduction and elimination rules for each connective and quantifier. In Hilbert calculus, in addition to the elimination rule, or modus ponens, some other axioms are given, both on → and on the connection of this with other connectives.

In propositional calculus, starting with a suitable reduced number of laws and rules, all of the tautologies are provable (completeness). The same is true for first-order logic, for which there are calculi in which all of the logically valid formulas are derivable; consequently the set of theorems is effectively enumerable. There is no complete logical calculus for second-order logic.

Intuitionistic logic has the same language as first-order (or second-order) logic; it is differentiated by the accepted laws. The *tertium non datur* $A \vee \neg A$, and the law of double negation $\neg\neg A \rightarrow A$ (in addition to some cases of De Morgan's laws, which depend on double negation) are rejected already at a propositional level. It also follows that some forms of reductio ad absurdum, which also depend on double negation, are not available. The case is similar for quantifiers where $\neg \forall x\, A(x) \rightarrow \exists x\, \neg A(x)$, which is logically valid in classical logic and a generalization of De Morgan's laws, is not accepted as an axiom: the reason is that in the case of an infinite domain, even if it can be proven that a property is valid for all of the elements, it does not follow that an element that falsifies the property can be identified.

First-order intuitionistic logic has its own semantics, derived from Kripke (1963), with respect to which it is complete.

11. The □ symbol indicates the end of a proof. □

References

Aigner, M., and G. M. Ziegler. 2006. *Proofs from the Book*. Berlin, Heidelberg: Springer-Verlag.

Albers, D. J., and C. Reid. 1987. "An Interview with Lipman Bers." *College Mathematics Journal* 18, no. 2 (January): 266–290.

Alexander, A. 2012. "From Voyagers to Martyrs." In Doxiadis and Mazur 2012, 1–51.

Angier, N. 2012. "The Mighty Mathematician You've Never Heard Of." *New York Times*, March 26, D4.

Archimedes. 2013. *Metodo*. Ed. F. Acerbi, C. Fontanari, and M. Guardini. Turin: Bollati Boringhieri.

Augustine of Hippo. 2001. *The Confessions*. Transl. Phillip Burton. New York: Alfred A. Knopf.

Auster, P. 2017. *4 3 2 1*. New York: Henry Holt.

Beccaria, G. L. 2016. *L'italiano che resta*. Turin: Einaudi.

Boole, G. 1847. *The Mathematical Analysis of Logic*. Cambridge, MA: Macmillan, Barclay & Macmillan.

Bricco, P. 2017. "Il fisico che gioca d'azzardo col tempo, conversazione con Carlo Rovelli." In *Il Sole–24 Ore* (Milan), May 28, 2017, 16.

Burdman, A., and S. Feferman. 2004. *Alfred Tarski: Life and Logic*. Cambridge: Cambridge University Press.

Cajori, F. (1929) 1993. *A History of Mathematical Notations*. 2 vols. bound as one. New York: Dover.

Calvino, I. 1995. *Saggi 1945–1985*. Ed. M. Barenghi. 2 vols. Milan: Mondadori.

Cantor, G. 1874. "Über eine Eigenschaft des Inbegriffes aller reellen algebraischen Zahlen." *Journal für die reine und angewandte Mathematik* (Crelle) 77: 258–262. English translation, "On a Property of the Set of Real Algebraic Numbers (*Cantor 1874*)," in Ewald 1996, 839–842.

Cantor, G. 1878. "Ein Beitrag zur Mannigfaltigkeitslehre [A contribution to the theory of manifolds]." *Journal für die reine und angewandte Mathematik* (Crelle) 84: 242–258.

Cantor, G. 1883. *Grundlagen einer allgemeinen Mannigfaltigkeitslehre. Eine mathematisch-philosophischer Versuch in der Lehre des Unendlichen.* Leipzig: Teubner. English translation, "Foundations of a General Theory of Manifolds: A Mathematico-Philosophical Investigation into the Theory of the Infinite (*Cantor 1883d*)," in Ewald 1996, 843–877.

Cantor, G. 1892. "Über eine elementare Frage der Mannigfaltigkeitslehre." *Jahresbericht der Deutschen Mathematiker Vereinigung* 1: 75–78. English translation, "On an Elementary Question in the Theory of Manifolds (*Cantor 1891*)," in Ewald 1996, 920–922.

Cantor, G., and R. Dedekind. 1937. *Cantor-Dedekind Briefwechsel.* Ed. E. Noether and J. Cavaillès. Paris: Hermann. English translation, "The Early Correspondence between Cantor and Dedekind" and "Cantor's Late Correspondence with Dedekind and Hilbert," in Ewald 1996, 843–877, 923–940.

Carmichael, R. D. 1930. *The Logic of Discovery.* Chicago: Open Court.

Carnap, R. 1932. "Die physikalische Sprache als Universalsprache der Wissenschaft." In *Erkenntnis*, vol. 2. English translation by M. Black in *The Unity of Science* (London: Kegan Paul, 1934).

Carnap, R. 1936. "Testability and Meaning." *Philosophy of Science* 3, no. 4: 419–471.

Cassirer, E. 1925. *Philosophie der symbolischen formen.* Berlin: Bruno Cassirer Verlag. English translation by R. Manheim, *The Philosophy of Symbolic Forms* (New Haven: Yale University Press, 1955).

Cauchy, A. L. (1821) 1897. "Cours d'analyse de l'École royale polytechnique." In *Oeuvres complètes*, series 2, vol. 3. Paris: Gauthier-Villars. English translation, *Cauchy's Cours d'analyse: An Annotated Translation*, ed. R. E. Bradley and C. E. Sandifer (New York: Springer, 2009).

Cauchy, A. L. (1823) 1899a. "Résumé des leçons données à l'École royale polytechnique sur le calcul infinitésimal." In *Oeuvres complètes*, series 2, vol. 4: 4–261. Paris: Gauthier-Villars.

Cauchy, A. L. (1829) 1899b. "Leçons sur le calcul différentiel." In *Oeuvres complètes*, series 2, vol. 4: 263–609. Paris: Gauthier-Villars.

Cauchy, A. L. 2013. *Hidden Harmony—Geometric Fantasies: The Rise of Complex Function Theory.* Transl. U. Bottazzini and J. Gray. Berlin: Springer.

Cavaillès, J. 1962. *Philosophie mathématique.* Paris: Hermann.

Cheng, E. 2017. *Beyond Infinity: An Expedition to the Outer Limits of Mathematics.* New York: Basic Books.

Darwin, C. (1887) 1958. *The Autobiography of Charles Darwin 1809–1882. With Original Omissions Restored. Edited with Appendix and Notes by his Granddaughter Nora Barlow.* London: Collins.

Dawkins, R. 2016. *The Ancestor's Tale: A Pilgrimage to the Dawn of Evolution.* Boston: Houghton Mifflin.

Dawson, J. W., Jr. 1997. *Logical Dilemmas: The Life and Work of Kurt Gödel.* Wellesley, MA: A. K. Peters.

De Millo, R. A., R. J. Lipton, and A. J. Perlis. 1979. "Social Processes and Proofs of Theorems and Programs." *Communications of the ACM* 22, no. 5 (January).

Dennett, D. C. 2017. *From Bacteria to Bach and Back.* New York: W. W. Norton.

de Santillana, G. 1985. *Fato antico e fato moderno.* Milan: Adelphi.

de Santillana, G., and H. von Dechend. 1977. *Hamlet's Mill: An Essay Investigating the Origins of Human Knowledge and Its Transmission through Myth.* Boston: Godine.

Devlin, K. 2000. *The Math Gene: How Mathematical Thinking Evolved and Why Numbers Are Like Gossip.* New York: Basic Books.

Douglas, M. 2007. *Thinking in Circles.* New Haven: Yale University Press.

Doxiadis, A. 2000. *Uncle Petros and Goldbach's Conjecture.* London: Faber and Faber.

Doxiadis, A. 2012. "A Streetcar Named (among Other Things) Proof: from Storytelling to Geometry, via Poetry and Rhetoric." In Doxiadis and Mazur 2012, 281–388.

Doxiadis, A., and B. Mazur, eds. 2012. *Circles Disturbed.* Princeton, NJ: Princeton University Press.

Einstein, A. 1935. "The Late Emmy Noether; Professor Einstein Writes in Appreciation of a Fellow-Mathematician." *New York Times*, May 4, 1935.

Einstein, A. 2011. *The Ultimate Quotable Einstein.* Ed. A. Calaprice. Princeton, NJ: Princeton University Press.

Euclid. 1956. *The Thirteen Books of Euclid's Elements.* Ed. T. L. Heath. Vol. 1. New York: Dover.

Euclid. 2003. *Dedomena: Euclid's Data, or, The Importance of Being Given.* Transl. C. M. Taisbak. Charlottenlund: Museum Tusculanum Press.

Euclid. 2014. *Tutte le opere.* Ed. F. Acerbi. Milan: Bompiani.

Euler, L. (1774) 1984. *Elements of Algebra.* Transl. J. Hewlett with additions by J.-L. Lagrange. New York: Springer. Translated from "Vollständige Anleitung zur nieder und höheren Algebra," in Euler, *Opera Omnia*, vol. I.1 (Leipzig: Teubner and O. Füssli, 1911).

Ewald, W. B., ed. 1996. *From Kant to Hilbert: A Source Book in the Foundations of Mathematics*, vol. 2. Oxford: Oxford University Press.

Frenkel, E. 2013. *Love and Math: The Heart of Hidden Reality.* New York: Basic Books.

Galilei, G. 1890–1909. *Le opere di Galileo Galilei.* 21 vols. Florence: Edizione Nazionale.

Galilei, G. (1638) 1958. *Discorsi e dimostrazioni matematiche intorno a due nuove scienze attinenti la meccanica e i moti locali.* Turin: Boringhieri.

Galilei, G. (1632) 1959. *Dialogo sopra i due massimi sistemi del mondo.* 2 vols. Milan: Biblioteca Universale Rizzoli.

Galilei, G. (1632) 1967. *Dialogue Concerning the Two Chief World Systems—Ptolemaic and Copernican.* Trans. S. Drake. Berkeley: University of California Press.

Gass, W. H. 1969. "Imaginary Borges." *New York Review of Books* 13, no. 9 (November 20, 1969), https://www.nybooks.com/articles/1969/11/20/imaginary-borges/.

Gould, S. J. 1996. "Why Darwin." A review of *Charles Darwin: Voyaging* by Janet Browne. *New York Review of Books* 43, no. 6 (April 4, 1996): https://www.nybooks.com/articles/1996/04/04/why-darwin/.

Grzegorczyk, A. 1964. "A Philosophically Plausible Formal Interpretation of Intuitionistic Logic." *Indagationes Mathematicae* 26: 596–601.

Haddon, M. 2003. *The Curious Incident of the Dog in the Nighttime*. London: Jonathan Cape.

Harari, Y. N. 2014. *Sapiens: A Brief History of Humankind*. London: Vintage Books.

Hardy, G. H. 1929. "Mathematical Proof." In Ewald 1996.

Hardy, G. H. (1940) 1967. *A Mathematician's Apology*. Preface by C. P. Snow. Cambridge: Cambridge University Press.

Heath, T. L. 1949. *The Mathematics in Aristotle*. Oxford: Oxford University Press.

Heath, T. L. (1921) 1981. *A History of Greek Mathematics*. 2 vols. New York: Dover.

Hilbert, D. 1900. "Mathematische Probleme." *Nachrichten von der Königlichen Gesellschaft der Wissenschaften zu Göttingen*: 253–297. English translation, "From Mathematical Problems (*Hilbert 1900b*)," in Ewald 1996, 1096–1105.

Hilbert, D. 1918. "Axiomatisches Denken." *Mathematische Annalen* 78: 405–415. English translation, "Axiomatic Thought (*Hilbert 1918*)," in Ewald 1996, 1105–1115.

Hilbert, D. 1922. "Neubegründung der Mathematik. Erste Mitteilung." *Abhandlungen aus dem mathematischen Seminar der Hamburgischen Universität* 1: 157–177. English translation, "The New Grounding of Mathematics: First Report (*Hilbert 1922a*)," in Ewald 1996, 1115–1134.

Hilbert, D. 1930. "Logic and the Knowledge of Nature." In Ewald 1996, 1157–1165.

Hilbert, D. 1983. "On the Infinite." In *Philosophy of Mathematics: Selected Readings*, 2nd ed., ed. P. Benacerraf and H. Putnam, 183–201. Cambridge: Cambridge University Press. Originally "Über das Unendliche (1925)," in *Mathematische Annalen* 95 (1926): 161–190.

Hilbert, D. 1992. *Natur und Mathematisches Erkennen*. Ed. D. E. Rowe. Lessons from 1919–1920 transcribed by P. Bernays. Basel: Birkhäuser.

Hilbert, D. 1996. "The Foundations of Mathematics." In *The Emergence of Logical Empiricism: From 1900 to the Vienna Circle*, ed. S. Sarkar, 228–244. New York: Garland. Originally "Die Grundlagen der Mathematik," *Abhandlungen aus dem mathematischen Seminar der Hamburgischen Universität* 6 (1927): 65–85.

Kac, M., and S. M. Ulam. (1968) 1992. *Mathematics and Logic*. New York: Dover.

Kenschaft, P. 1988. "An Interview with Marguerite Lehr: In Memoriam." *Association for Women in Mathematics*, newsletter 18, no. 2 (March-April): 9–11.

Kripke, S. 1963. "Semantical Considerations on Modal Logic." *Acta Philosophica Fennica* 16: 83–94.

Leibniz, G. W. 1969. *Gottfried Wilhelm Leibniz: Philosophical Papers and Letters*. Transl. L. E. Loemker. Dordrecht: D. Reidel. Originally *Leibnizens mathematische Schriften*, ed. C. I. Gerhardt, 7 vols. Halle: Schmidt, 1849–1863; reprinted, Hildesheim: Georg Olms, 1961.

Lloyd, G. E. R. 2012. "Mathematics and Narrative: An Aristotelian Perspective." In Doxiadis and Mazur 2012, 389–406.

Lolli, G. 2016a. *Tavoli, sedie, boccali di birra*. Milan: Raffaello Cortina.

Lolli, G. 2016b. "Le 'Lezioni' perdute di Calvino." In *Matematica e letteratura. Analogie e convergenze*, ed. P. Maroscia, C. Toffalori, F. S. Tortoriello, and G. Vincenzi, 53–79. Turin: Utet.

Lolli, G. 2017. *Ambiguità. Un viaggio tra letteratura e matematica*. Bologna: Il Mulino.

Manzoni, A. (1840) 2014. "Storia della colonna infame." In *I promessi sposi*. Milan: Biblioteca Universale Rizzoli.

Manzotti, R., and T. Parks. 2016–2017. "On Consciousness: A Series of Conversations between Riccardo Manzotti and Tim Parks." *New York Review of Books* (blog), December 8, 2016, December 30, 2016, January 26, 2017, https://www.nybooks.com/topics/on-consciousness/.

Mazur, B. 2003. *Imagining Numbers: (Particularly the Square Root of Minus Fifteen)*. New York: Farrar, Straus and Giroux.

Mazur, B. 2012. "Visions, Dreams, and Mathematics." In Doxiadis and Mazur 2012, 183–210.

McCulloch, W., and W. Pitts. 1943. "A Logical Calculus of Ideas Immanent in Nervous Activity." *Bulletin of Mathematical Biophysics* 5: 115–133.

Mirzakhani, M. 2014. "Interview." *The Guardian*, August 13, https://www.theguardian.com/science/2014/aug/13/interview-maryam-mirzakhani-fields-medal-winner-mathematician.

Musil, R. (1913) 1990. "The Mathematical Man." In *Precision and Soul: Essays and Addresses*, ed. and transl. B. Pike and D. S. Luft, 39–43 Chicago: University of Chicago Press.

Neumann, J. von. (1966) 2004. "The Mathematician." In *Musings of the Masters*, ed. R. G. Ayoub, 169–184. Washington: Mathematical Association of America.

Osen, L. M., ed. 1974. *Women in Mathematics*. Cambridge, MA: MIT Press.

Pacchioni, G. 2017. *Scienza, quo vadis?* Bologna: Il Mulino.

Pasch, M. 1882. *Vorlesungen über neuere Geometrie*. Leipzig: Teubner.

Peacock, G. 1833. "Report on the Recent Progress and Present State of Certain Branches of Analysis." In *Report of the Third Meeting of the British Association for the Advancement of Science; Held at Cambridge in 1833*, 185–352. London: John Murray.

Plato. 1977. *Plato*, vol. 2. Transl. W. R. M. Lamb. London: William Heinemann.

Plutarch. 1955. *Plutarch's Lives*. Vol. 5. Transl. Bernadotte Perrin. London: William Heinemann.

Romano, L. 1998. *L'eterno presente*. Turin: Einaudi.

Rosen, C. 2012. "Freedom and Art." *New York Review of Books* 59, no. 8, May 10, 2012. https://www.nybooks.com/articles/2012/05/10/freedom-and-art/.

Roth, P. 2010. *Novels 1993–1995*. New York: Library of America.

Rovelli, C. 2018. *The Order of Time*. Transl. E. Segre and S. Carnell. New York: Riverhead Books.

Russell, B. 1901. "Recent Work on the Principles of Mathematics." *International Monthly* 4 (July 1901): 83–101. Reprinted in 1918 as "Mathematics and the Metaphysicians," in *Mysticism and Logic and Other Essays*, 74–96. London: Longmans, Green.

Russell, B. 1903. *The Principles of Mathematics*. London: Allen and Unwin.

Russell, B. 1906. "Les paradoxes de la logique." *Revue de Métaphysique et de Morale* 14: 627–650. English translation, "On 'Insolubilia' and Their Solution by Symbolic Logic," in *Essays in Analysis* (London: Allen and Unwin, 1973), 190–214.

Russell, B. 1918. *Mysticism and Logic*. London: Longmans, Green.

Russell, B. 1973. *Essays in Analysis*. Ed. D. Lackey. London: George Allen and Unwin.

Russo, L., G. Pirro, and E. Salciccia. 2017. *Euclide: il I libro degli Elementi*. Rome: Carocci.

Schmalz, R., ed. 1993. *Out of the Mouths of Mathematicians*. Washington: MAA Press.

Shapin, S. 2008. *The Scientific Life: A Moral History of a Late Modern Vocation*. Chicago: University of Chicago Press.

Smith, J. E. H. 2016. *The Philosopher: A History in Six Types*. Princeton, NJ: Princeton University Press.

Stevin, S. 1958. *The Principal Works of Simon Stevin*. Vol. II, *Mathematics*. Ed. D. J. Strujk. Amsterdam: C. V. Swets & Zeitlinger.

Struyk, D. J. 1981. "Bolyai, János Johann." In *Dictionary of Scientific Biography*, ed. C. Gillespie. New York: Charles Scribner's Sons.

Szabó, Á. 1978. *The Beginnings of Greek Mathematics*. Dordrecht: D. Reidel.

Tabucchi, A. 2013. "Storie che non sono la mia." *Il Sole–24 Ore*, September 22, 2013, 36.

Thurston, W. P. 1994. "On Proof and Progress in Mathematics." *Bulletin of the American Mathematical Society* 30, no. 2:161–177.

Torricelli, E. 1919. *Opere di Evangelista Torricelli*. Ed. G. Loria and G. Vassura. 4 vols. Faenza: Typo-Litografia G. Montanari.

van der Waerden, B. L. 1950. *Erwachende Wissenschaft*. Groningen: P. Noordhoff.

van Heijenoort, J., ed. 1967. *From Frege to Gödel*. Cambridge, MA: Harvard University Press.

Varvaro, A. 2017. *Il fantastico nella letteratura medievale*. Bologna: Il Mulino.

Vernant, J.-P. 1962. *Les origines de la pensée grecque*. Paris: PUF. English translation, *The Origins of Greek Thought* (Ithaca, NY: Cornell University Press, 1984).

Villani, C. 2015. *Birth of a Theorem: A Mathematical Adventure*. New York: Farrar, Straus and Giroux.

Weinberg, A. M. 1967. *Reflections on Big Science*. Cambridge, MA: MIT Press.

Index of Names